Success guides

Leckie ✕ Leckie

Intermediate 2
Biology

✕ Andrew Morton ✕

Contents

Unit 1 – Living cells

Unit 2 – Environmental biology and genetics

Unit 3 – Animal physiology

Exam skills and support

Cells

- Living things are called organisms
- All organisms are made of cells
- Some organisms, like bacteria, consist of only one cell. They are unicellular
- Some, like us, are made of many millions of cells. We are multi-cellular

Cell Size

Cells are usually only visible under a microscope.

They are much smaller than a millimetre, so we measure them in microns (μm).

There are a thousand microns in a millimetre.

nerve cell 100 μm diameter

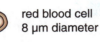

liver cell
20 μm diameter

red blood cell
8 μm diameter

bacterium
1 μm diameter

A human cheek cell

Cheek cells make up the epithelium (lining) of the cheek.

Nucleus – controls the manufacture of protein by the cell. it contains chromosomes which are made of DNA.

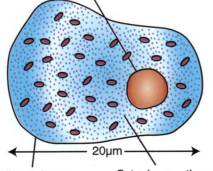

20μm

Cell membrane – acts as a boundary and controls what enters and leaves the cell.

Cytoplasm – the contents of the cell, contained by the cell membrane. Full of tiny structures and thousands of different chemicals.

A leaf cell

These cells can be found in the mesophyll (middle) of a plant leaf.

Top-Tip
Make sure you know the functions of the parts of a cell.

Chloroplasts – green structures which carry out photosynthesis

Nucleus

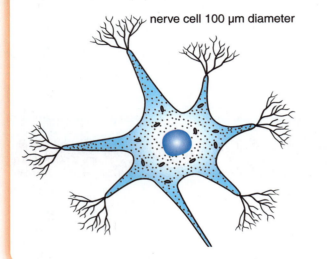

30μm

Cell wall – made of cellulose, it is permeable to water and is a tough protective coating of the cell.

Vacuole – filled with water and dissolved substances it gives support and firmness to the cell.

A yeast cell

Yeast is a unicellular fungus.

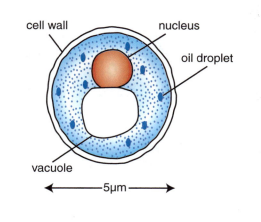

A bacterium

Bacteria are primitive cells which appeared on Earth long before other types of cells.

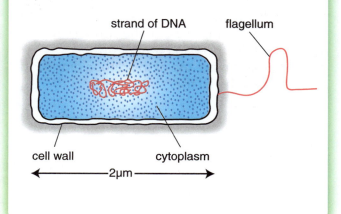

Animal and plant cells

Similarities and Differences

They both have:	Only plant cells have:
a nucleus	a cell wall
cytoplasm	vacuoles
a cell membrane	chloroplasts

There are a few exceptions to the information in the table above.

For example, red blood cells have no nucleus, and small vacuoles are found in some animal cells.

Top-Tip
Make sure you know the differences between animal and plant cells.

Quick Test

1. Name two cell structures which are found only in plant cells.
2. What is the function of the cell membrane?
3. What is the actual size of the cell shown opposite?
4. What important process takes place in chloroplasts?
5. What important substance is found in the nucleus of a cell?
6. Do bacteria have a nucleus?
7. Where is the cell membrane found in a plant cell?

Cell is magnified 400x

←20mm→

Answers 1. chloroplast and cell wall 2. to control the entry and exit of substances 3. 50μm 4. photosynthesis 5. DNA 6. no 7. pressed against the inside cell wall

Useful cells 1

Key Facts

- Humans have made use of unicellular organisms for thousands of years.
- Yeast produces alcohol and carbon dioxide during the fermentation of sugar.
- Brewers make use of the alcohol when making beer and wine.
- Bakers make use of the carbon dioxide to make bread rise.

Yeast cells reproducing by budding

Making alcohol

When **yeast** is mixed with sugar and water in the absence of air it produces **alcohol** and **carbon dioxide**. This process is known as **fermentation**. The yeast does this to obtain energy for itself. Yeast cannot **photosynthesise** because it is a **fungus** and therefore has no **chlorophyll**. It needs sugar for energy, just as we do.

sugar + yeast → alcohol + carbon dioxide + energy

The flavours in different alcoholic drinks result from the different ingredients used to make the drink, e.g. grapes are used to make wine; barley and hops are used to make beer.

If yeast is provided with a supply of air or oxygen it will produce water instead of alcohol. It is therefore important that it is kept in **anaerobic** (oxygen-free) conditions during **brewing**.

Top Tip
Remember, when yeast respires without oxygen it produces alcohol and carbon dioxide.

Top Tip
It's worth remembering that the proper chemical name for alcohol is ethanol.

air lock (to allow carbon dioxide to escape, but to keep out unwanted microbes)

carbon dioxide

sugar + water + yeast + minerals

The fermenter is kept warm to encourage the yeast to grow.

Making bread

When yeast is mixed with flour, sugar and water in a dough, it ferments the sugars in the dough and produces **alcohol** (ethanol) and **carbon dioxide**. The carbon dioxide is trapped as bubbles in the dough and makes it expand (rise). When the dough has risen it is baked at a high temperature in an oven. The baking process kills the yeast, drives off the alcohol and firms the bread.

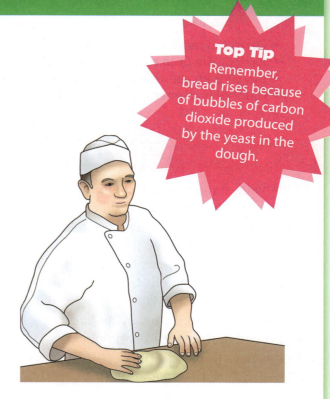

Top Tip
Remember, bread rises because of bubbles of carbon dioxide produced by the yeast in the dough.

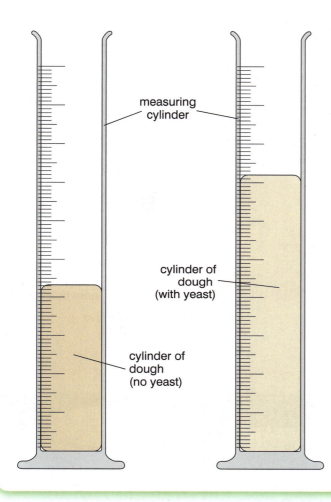

measuring cylinder

cylinder of dough (with yeast)

cylinder of dough (no yeast)

The effect of yeast on dough can be investigated in the lab using measuring cylinders. Dough can be put in the cylinders and left to rise. The distance the dough moves up the cylinder indicates the level of activity of the yeast.

Dough without yeast does not rise, showing that it is the yeast which causes the dough to rise.

Set up an experiment where the cylinders are all at different temperatures to check the effect of temperature on the activity of the yeast.

Quick Test

1. What kind of organism is yeast?

2. What important process are fungi unable to carry out?

3. What name is given to the process in which yeast turns sugar into alcohol in the absence of air?

4. If the yeast is provided with oxygen what will it produce instead of alcohol?

5. Name the gas that causes bread to rise.

Answers 1. A fungus. 2. Photosynthesis. 3. Fermentation. 4. Water. 5. Carbon dioxide.

Useful cells 2

- Fungi can produce a wide range of antibiotics that can be used to kill bacteria.
- Bacteria are used to make yoghurt.
- Biogas is an energy-rich fuel that is produced with the help of bacteria.
- Gasohol is an energy-rich fuel that is produced with the help of yeast.

Antibiotic production

In the wild, fungi produce chemicals and release them into the soil to kill competitors like bacteria. These chemicals are called **antibiotics** and have been used by humans for over 60 years to treat bacterial infections. Unfortunately, bacteria evolve very quickly and can adapt to new hostile environments. This means many bacteria are becoming resistant to antibiotics. For this reason, the use of antibiotics is restricted and doctors will not always prescribe antibiotics when you are ill. Also, many diseases are caused by viruses, and antibiotics have no effect on viruses.

Top Tip
Some bacteria have become resistant to antibiotics because we use antibiotics too much.

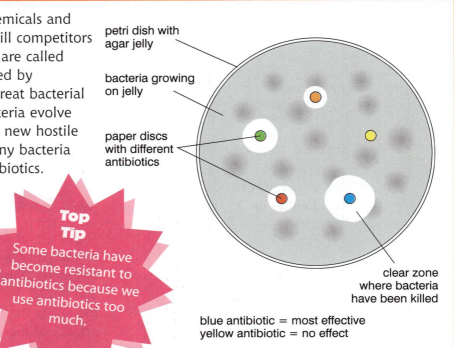

petri dish with agar jelly

bacteria growing on jelly

paper discs with different antibiotics

clear zone where bacteria have been killed

blue antibiotic = most effective
yellow antibiotic = no effect

Yoghurt production

Milk contains a sugar called **lactose**. Some bacteria can convert lactose into **lactic acid**, causing curdling of the milk. This is what happens when milk is left too long and goes sour. However, with certain species of bacteria the soured milk is more edible and we call it **yoghurt**. The yoghurt is more stable and lasts longer than milk because of the drop in **pH** caused by the lactic acid. (pH is a measure of the acidity or alkalinity of a solution. A pH of 7 is neutral. A pH less than 7 is acid and a pH higher than 7 is alkaline.)

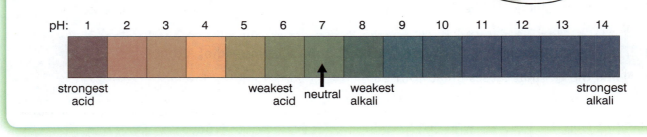

pH: 1 2 3 4 5 6 7 8 9 10 11 12 13 14

strongest acid   weakest acid   neutral   weakest alkali   strongest alkali

Alternative fuel production

Some bacteria produce **methane** gas in conditions where oxygen concentrations are low. This gas is rich in chemical energy and can be burned. It is sometimes called **biogas**. Biogas can be produced from organic waste and is a gas commonly produced by decaying rubbish. For this reason it is particularly useful because it can be produced from a widely available raw material. The gas supplied to our homes is mostly methane that has been produced by the action of bacteria on the decaying remains of plants and animals millions of years ago.

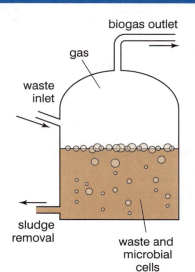

biogas outlet

gas

waste inlet

sludge removal

waste and microbial cells

Top Tip
Biogas and gasohol are very important because they are renewable sources of energy.

Petrol is made from oil, a resource which is rapidly running out. So, humans need alternative supplies of energy that are renewable, i.e. that won't run out. Alcohol is such a fuel, which can be made by the action of yeast on sugar. Alcohol can be used either on its own or mixed with petrol to run car engines. The alcohol-petrol mixture is often called **gasohol**.

Quick Test

1. Name the organisms that produce antibiotics.
2. What happens to bacteria if antibiotics are used too often to kill them?
3. Name the sugar found in milk.
4. Which organisms are used to make yoghurt?
5. What is meant by a renewable source of energy?

Answers 1. Fungi. **2.** They become resistant to the antibiotics. **3.** Lactose. **4.** Bacteria. **5.** A source of energy that never runs out.

Diffusion

Key Facts

- Diffusion is the naturally occurring movement of gas or liquid molecules from areas of high concentration to areas of low concentration.
- Diffusion is a process used by all organisms to transport substances over very short distances.

Concentration gradients

When molecules in gases or liquids are unevenly distributed we say that a **concentration gradient** exists. However, because the molecules are constantly moving around they tend to distribute themselves evenly, i.e. they always move from an area of high concentration to an area of low concentration, down the concentration gradient. This process is called **diffusion**.

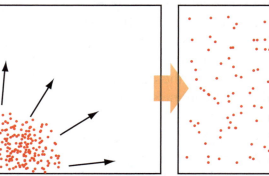
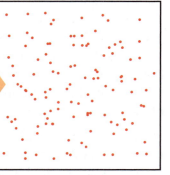

The importance of diffusion

Diffusion is a very important process for living things because it is the way in which substances are often moved about from cell to cell and to their surroundings. Diffusion is a very slow process. Consequently, many organisms have had to adapt in a wide variety of ways to cope with this problem.

If we waited for diffusion to get a molecule of oxygen from our mouth to our big toe we would have to wait many years. So, we use our breathing muscles to bring air into our lungs and we use our heart to pump oxygenated blood to our feet.

Oxygen moves from the air sacs in our lungs into the red blood cells by diffusion. This process is so slow that we have a huge surface area in our lungs to allow it to happen. In fact, the surface area of our lungs is about the same as the surface area of the walls of an average classroom. Similarly, our intestines have a huge surface area for the absorption of digested food.

In plants also, diffusion is the process by which the gases oxygen and carbon dioxide enter and leave the leaves.

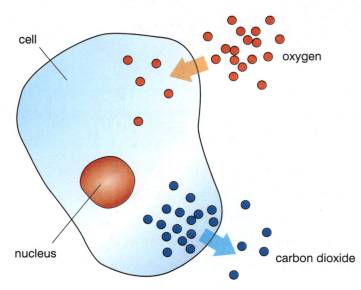

Cell membranes and diffusion

Substances have to move into or out of cells through the **cell membrane**.
If the molecules or atoms are small enough, they can pass through tiny pores in the membrane, but if they are larger they can only pass through slowly or not at all. For this reason we say cell membranes are semi or **selectively permeable**.

Hypertonic and hypotonic solutions

Biologists use special terms to compare solutions that have different strengths.

If one solution has more substances dissolved in it than another it is said to be **hypertonic** to the other solution. The weaker solution is described as **hypotonic** to the stronger solution.

If both solutions are the same strength they are said to be **isotonic**. Many drinks sold to athletes are said to be isotonic, so that they do not upset the athlete's water balance just before or during an endurance event.

Water is always hypotonic to other solutions which are hypertonic to the pure water, e.g. The body fluids of a freshwater fish will be hypertonic to the water in which they swim.

Top Tip
Make up a mnemonic to help you remember the difference between hypotonic and hypertonic.

Quick Test

1. Name two important substances that might enter a cell by diffusion.

2. What term is used to describe membranes because they only allow certain substances to pass through them?

3. In the diagram opposite what term can be used to describe solution A in contrast to solution B?

4. If a cell is placed in fresh water, what term could be used to describe the fresh water in contrast to the cell contents?

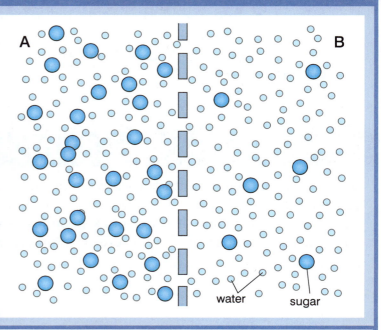

A   B

water   sugar

Osmosis

- The diffusion of water through a selectively-permeable membrane from an area of high concentration of water molecules to an area of low concentration of water molecules is called osmosis.
- Plant and animal cells behave differently in water because plant cells have a protective cell wall, which is absent in animal cells.

The diffusion of water

Any substance dissolved in water has the capability of diffusing. However, water molecules will also diffuse down a concentration gradient. So, if a strong sugar solution is separated from a weak sugar solution by a selectively-permeable membrane, the water, which can pass easily through the membrane, will tend to diffuse from the weak solution to the strong solution. This movement of the water molecules is called **osmosis**.

Top Tip
Don't forget that the diffusion of water is so important it is given a special name – osmosis.

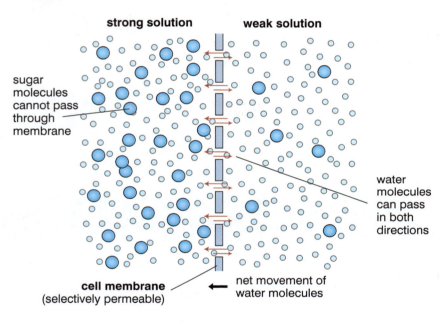

strong solution   weak solution

sugar molecules cannot pass through membrane

water molecules can pass in both directions

cell membrane
(selectively permeable)

← net movement of water molecules

Osmosis in the kitchen

We make use of osmosis in the kitchen when preparing a salad. If we wash lettuce leaves in fresh water, the water flows into the plant cells by osmosis and keeps them crisp and firm. If salt is sprinkled on the leaves, they go limp. Just what is happening at cellular level is explained on the next page.

Osmotic effects on animal and plant cells

Fresh water has the highest concentration of water molecules. This means water will always diffuse from fresh water towards a solution of salts or sugars in water. If you put a cell in fresh water the cell will gain water, whereas if you put a cell in a strong solution of salt or sugar, it will lose water.

Animal cells react differently to plant cells because plant cells have a **cell wall** and animal cells do not. The table below summarises what happens to animal and plant cells in different circumstances.

	Fresh water	Salt water
Animal cell	Swells up and bursts	Shrinks
Plant cell	Swells up and becomes firm	Shrinks and becomes limp

We use special words to describe plant cells that are firm and plant cells that are limp. Firm cells are **turgid**, and limp cells are **flaccid.**

In fact, if a plant cell is put in a very strong salt or sugar solution it will lose so much water that its cytoplasm and membrane will shrink away from the cell wall. It is then said to be **plasmolysed** and the process is called plasmolysis.

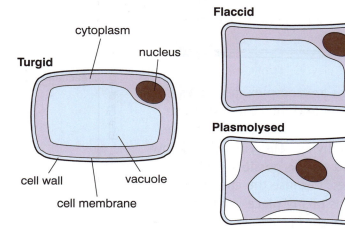

Quick Test

Three identical plant cells were treated as follows: one was placed in fresh water and the other two in different salt solutions. They were left for 20 minutes then examined under the microscope. Their appearance is shown below.

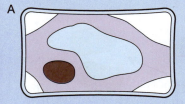
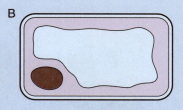
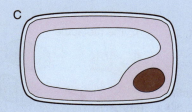

1. Which cell was placed in fresh water?

2. What term describes the appearance of cell C?

3. What term describes the appearance of cell A?

4. Which cell was placed in a strongly hypertonic solution?

Answers 1. C. 2. Turgid. 3. Plasmolysed. 4. A.

Enzymes

- Enzymes are biological catalysts made by all living cells.
- Enzymes are highly specific, i.e. each enzyme will only catalyse one reaction.

Properties of enzymes

Enzymes are **catalysts**. Catalysts are substances that speed up chemical reactions without being changed themselves. They do this by lowering the energy input that is required to start a chemical reaction. This energy input is called the **activation energy**. A simple analogy is the energy needed to strike a match. Once that energy has been expended, the chemical reactions proceed on their own and the match burns.

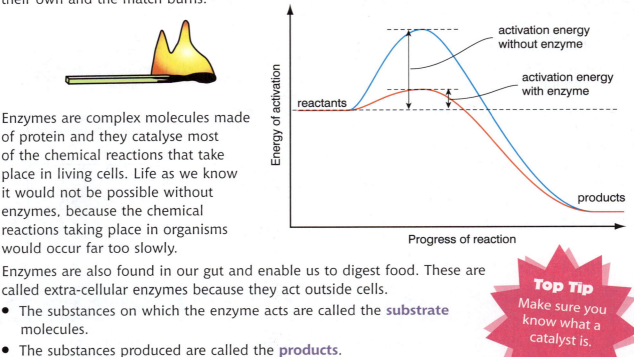

Enzymes are complex molecules made of protein and they catalyse most of the chemical reactions that take place in living cells. Life as we know it would not be possible without enzymes, because the chemical reactions taking place in organisms would occur far too slowly.

Enzymes are also found in our gut and enable us to digest food. These are called extra-cellular enzymes because they act outside cells.

- The substances on which the enzyme acts are called the **substrate** molecules.
- The substances produced are called the **products**.
- The substrates combine briefly with the enzyme at a particular site called the **active site**.

Top Tip
Make sure you know what a catalyst is.

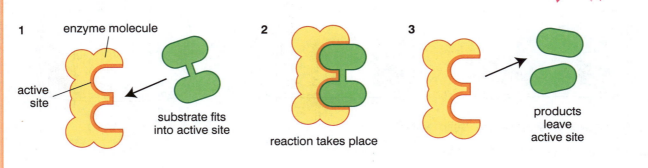

1 enzyme molecule

active site

substrate fits into active site

2 reaction takes place

3 products leave active site

Enzyme specificity

There are hundreds of different enzymes in cells, because each enzyme can only catalyse one reaction. This is because the active sites only allow one **specific** substrate to fit; a bit like a key in a lock.

Top Tip
Don't forget that cells are full of hundreds of enzymes all catalysing different reactions.

Degradation and synthesis

All the chemical reactions taking place in an organism are described as its **metabolism**. These chemical reactions are of two types:

- **Synthesis (build-up) reactions** in which complex molecules are built from simpler molecules. For example, amino acids are combined together to make proteins.

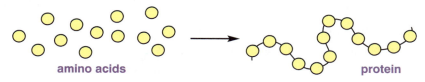

amino acids   protein

- **Degradation (breakdown) reactions** in which complex molecules are broken down to simpler molecules. For example, starch can be broken down to maltose or glucose.

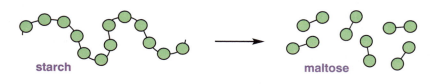

starch   maltose

Enzymes catalyse both types of reaction.

Quick Test

1. Describe two features of a catalyst.

2. What effect do enzymes have on activation energy?

3. Where on the enzyme molecule does the chemical reaction take place?

4. The enzyme pepsin digests protein in the stomach. What kind of reaction is this?

5. What is meant by the term 'metabolism'?

6. How many glucose molecules are found in one molecule of maltose?

Answers 1. They speed up chemical reactions and they are unchanged during the process. 2. They lower it. 3. The active site. 4. Degradation/breakdown. 5. The sum total of all chemical reactions in an organism. 6. 2.

Enzyme activity

Key Facts
- Enzymes are affected by the temperature and pH of their surroundings.
- Enzymes are denatured by high temperatures and by strong acids and alkalis.

Factors affecting enzyme activity

Because enzymes are made of protein their shape can be altered easily when their surroundings change. If the shape of the active site is altered then the enzyme will no longer be able to function. Many poisons stop enzyme activity by altering the shape of the active site. Changes to temperature and pH also affect the activity of enzymes.

The effect of temperature

Enzymes are very sensitive to temperature and work best at an ideal temperature called the **optimum** temperature. As the temperature drops below the optimum, the activity of the enzyme decreases because the substrate molecules do not move so quickly.

As the temperature rises above the optimum, the activity of the enzyme decreases because the molecular structure of the enzyme is affected by the increased movement of its own atoms. The vibration of the atoms alters the shape of the active site permanently and we say that the enzyme has been **denatured**.

Top Tip
Remember that high temperatures denature enzymes.

The optimum temperature of enzymes working in our bodies is around 40°C and this is the reason we maintain a constant body temperature like other mammals and birds. Other enzymes have optimum temperatures both lower and higher than 40°C. For example, the enzymes found in Arctic fish have a lower optimum temperature, and the enzymes found in the algae of hot springs have a higher optimum temperature.

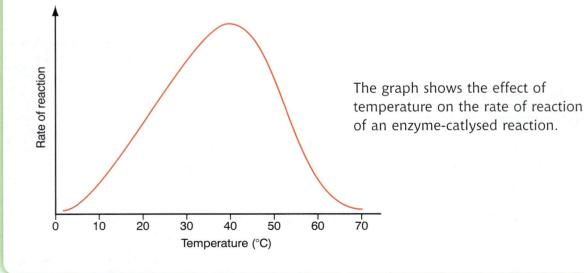

The graph shows the effect of temperature on the rate of reaction of an enzyme-catlysed reaction.

The effect of pH

Enzymes are affected by changes to pH for similar reasons to those described opposite. Changes in pH away from the optimum alter the shape of the active site and denature the enzyme.

Most enzymes work best at a pH close to neutral (pH7), but there are a few exceptions. **Pepsin**, which is found in our stomach, works best at pH2.

Like temperature, the pH of our blood and tissue fluid is kept constant to maintain optimum conditions for enzyme activity.

Top Tip
Remember that the best conditions for enzymes are described as the optimum conditions.

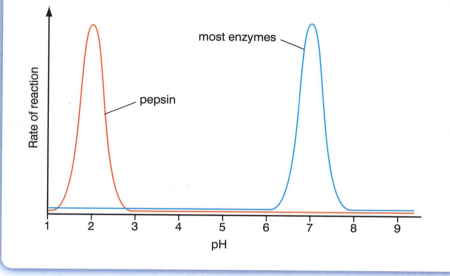

Quick Test

The graph shows the activity of an enzyme through a range of different pH values.

1. What is the optimum pH for the enzyme?

2. How has the activity of the enzyme been measured?

3. What term is used to describe an enzyme damaged by an unsuitable pH?

4. Predict the activity of the enzyme if the experiment had been carried out at 70°C.

Answers 1. pH6. 2. By measuring the volume of gas given off in one minute. 3. Denatured. 4. The enzyme would be inactive.

Experiments with enzymes

Key Facts

When carrying out experiments:
- repetition improves reliability
- only one variable should be altered
- it is important to set up controls for comparison.

An experiment with amylase

Amylase is an enzyme that catalyses the breakdown of **starch** (the substrate) to **maltose** (the product). Maltose is a sugar that can be broken down to **glucose**. We have amylases in our digestive system to break down foods like bread and potatoes. Even our saliva contains amylase.

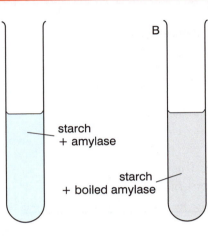

A — starch + amylase

B — starch + boiled amylase

Top Tip
This is a very simple introductory experiment, but it is very important you understand the need to control variables.

Set up two test-tubes with starch solution in each. Add amylase solution to one and boiled amylase solution to the other. Leave them for 1 hour then test both for starch and sugar. The starch will disappear in test-tube A but not in B. Sugar will appear in test-tube A but not in B.

When carrying out experiments you should record your work under the following headings:

AIM, METHOD, RESULTS, CONCLUSION.

An experiment with catalase

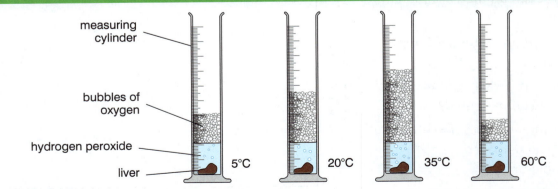

measuring cylinder

bubbles of oxygen

hydrogen peroxide

liver

5°C   20°C   35°C   60°C

Catalase is an enzyme found in many cells, but particularly in liver cells. It catalyses the breakdown of hydrogen peroxide (a poisonous end-product of cell metabolism) to oxygen and water.

Hydrogen peroxide $\xrightarrow{\text{catalase}}$ water + oxygen

The activity of catalase can be measured by recording the volume of bubbles of oxygen produced when a piece of liver is added to a dilute solution of hydrogen peroxide. **Take care, hydrogen peroxide is dangerous, so you should wear goggles when carrying out this experiment.**

Set up four measuring cylinders with hydrogen peroxide in each, all at different temperatures. Add a drop of detergent to each so that the bubbles of oxygen make a froth that can be measured. Add a piece of liver to each cylinder and record the volume of froth produced after five minutes.

To make the results more **reliable**, the experiment should be repeated a number of times and the results averaged. This reduces the effect of unusual results.

An experiment with phosphorylase

Phosphorylase is an enzyme found in potatoes. It catalyses the synthesis (build-up) of starch from sugar. The sugar substrate is a form of glucose called glucose-1-phosphate. The enzyme can be obtained from potato tissue by mashing raw potato and filtering the juice.

Add the enzyme extract to glucose-1-phosphate solution and check to see whether starch is produced. If starch is produced you must demonstrate that this could not happen on its own without the potato extract. This is where a **control** experiment is needed. In this, every detail is the same except that there is no enzyme. If no starch is produced, then this proves that the enzyme is causing the change.

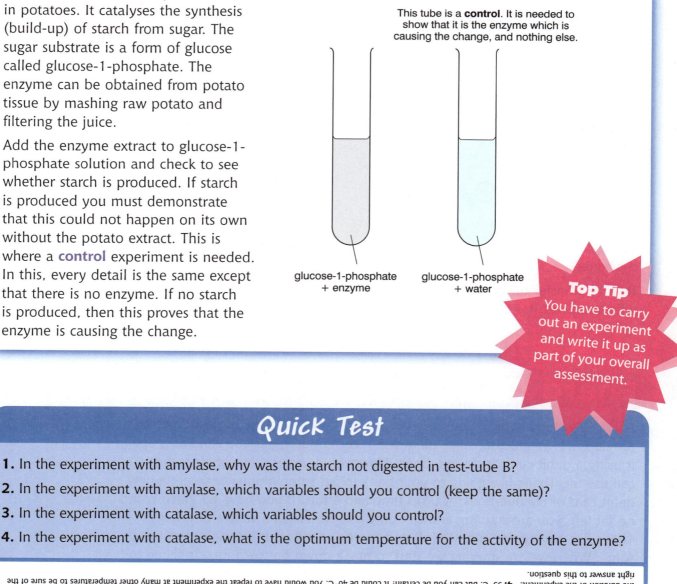

This tube is a **control**. It is needed to show that it is the enzyme which is causing the change, and nothing else.

glucose-1-phosphate + enzyme

glucose-1-phosphate + water

Top Tip
You have to carry out an experiment and write it up as part of your overall assessment.

Quick Test

1. In the experiment with amylase, why was the starch not digested in test-tube B?

2. In the experiment with amylase, which variables should you control (keep the same)?

3. In the experiment with catalase, which variables should you control?

4. In the experiment with catalase, what is the optimum temperature for the activity of the enzyme?

Energy release

- All organisms obtain useful energy by respiration.
- Energy-rich compounds like glucose are broken down to provide energy for the synthesis of ATP.
- ATP is the universal source of immediate energy for chemical reactions.

Respiration

All organisms need energy to survive, and they obtain energy from energy-rich compounds like glucose and fat by a process called **respiration**. This is a chemical process.

Respiration takes place in every living cell in a series of controlled steps, each catalysed by an enzyme. Usually, oxygen is required, but it is possible for the process to take place in the absence of oxygen.

The energy produced is used to make an energy-rich compound called **Adenosine TriPhosphate (ATP)**. The energy is used to add a phosphate group (P_i) to **Adenosine diphosphate (ADP)**:

$$ADP + P_i + energy \rightarrow ATP$$

Top Tip
Respiration takes place in all living cells, including plants, all the time.

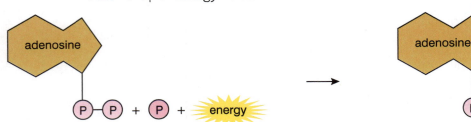

When ATP passes on its energy it does so by the transfer of the third phosphate molecule.

It is the transfer of this phosphate molecule which energises the compound which receives it.

Energy currency

ATP is like ready cash in your pocket. Glucose can be readily broken down to provide ATP. So glucose is like money in a current account: you can easily spend it using a debit card or cheque-book. However, if you run out of glucose you can use fat. That's like going to your bank to get some money or to get a new debit card or cheque-book: it takes time and effort. If you have no glucose and no fat left, you are starving and can only use protein as a source of energy. That's like having no money at all and having to sell your belongings to raise some. When we are starving we start to break down our own muscle tissue to make ATP.

The role of ATP

ATP is often referred to as the universal energy currency. It is used by all living things any time energy is required. It is not used as an energy store, but as an immediate source of energy, so that the rate of synthesis of ATP always matches the rate of use. Humans contain around 50 g of ATP whether asleep or running the 100 metres.

ATP is used as a source of energy for many chemical reactions taking place in cells and for much else besides. For example it is used to:

- enable sperm to swim
- make cilia flick to and fro
- cause muscles to contract
- permit the cell membrane to engulf particles and liquids
- force molecules up a concentration gradient
- pull chromatids apart during cell division
- move materials about the cell
- transmit nerve impulses.

Top Tip
Think of ATP as cash-in-hand, and glucose and fats as money in the bank.

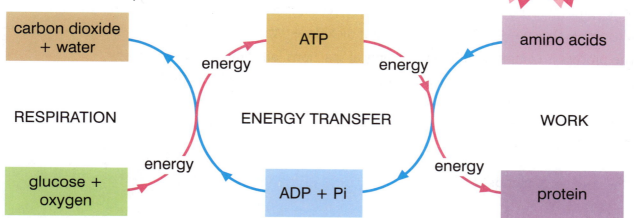

Ultimately, almost all the energy is released as heat, and mammals and birds make great use of this heat energy to maintain a constant body temperature. See pages 82 and 83.

Quick Test

1. What compounds are common sources of energy for respiration?
2. Name the molecule that is regarded as the universal energy currency.
3. What part of this molecule transfers energy from one compound to another?
4. Describe three uses of ATP within a cell.
5. What is the ultimate fate of all energy used in a cell?

Answers 1. Glucose and fat. 2. ATP. 3. The phosphate group. 4. Muscle contraction, transmission of nerve impulses, movement of materials round the cell. 5. It ends up as heat energy.

Aerobic and anaerobic respiration

Key Facts
- Aerobic respiration is much more efficient than anaerobic respiration
- The end products of aerobic respiration are carbon dioxide and water
- The end products of anaerobic respiration are carbon dioxide and ethanol in plants, and lactic acid in animals

Respiration with oxygen

Respiration with oxygen is called **aerobic respiration**. Aerobic respiration is by far the most efficient form of respiration. One glucose molecule has enough chemical energy in it for the production of 38 molecules of ATP, and fats contain roughly two times more energy than glucose. Glucose is the most common respiratory substrate, and in aerobic conditions the end-products are water and carbon dioxide, both of which are very low in energy.

Glucose + Oxygen + ADP + P_i → Carbon dioxide + Water + ATP

The process is complex, with many chemical reactions, all catalysed by enzymes.

The first stage results in the break-up of a glucose molecule into two molecules of **pyruvic acid**. This process is called **glycolysis**. In this stage only two molecules of ATP are produced.

GLYCOLYSIS

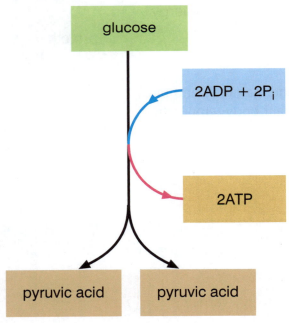

Top Tip
Remember the total number of ATP molecules produced from one molecule of glucose in aerobic respiration is 38.

In the following stage, the pyruvic acid is broken down further to obtain hydrogen, which eventually combines with oxygen to form water. During this second stage, carbon dioxide is released. It is during the formation of water that most of the ATP is produced; 36 molecules in all.

Respiration without oxygen

Some cells can respire in the absence of oxygen. This is called **anaerobic respiration**.

Anaerobic respiration is much less efficient than aerobic respiration. This is because the second stage cannot take place, i.e. the pyruvic acid produced during glycolysis cannot be converted to carbon dioxide and water because there is no oxygen to combine with the hydrogen to make water. Consequently, only two molecules of ATP are produced and the end products of the process are still high in energy. In plants the high-energy end product is ethanol, in animals it is lactic acid. Plants also produce carbon dioxide during anaerobic respiration.

Both ethanol (alcohol) and lactic acid are poisons. Fortunately, animals can convert the lactic acid into non-toxic compounds with the help of oxygen. So, after strenuous activity we will breathe heavily for a time, repaying the **oxygen debt**. Fitness training improves our ability to tolerate lactic acid and to remove it after exercise.

On the other hand, plants and fungi cannot remove the ethanol, and if they continue to respire anaerobically, will eventually die. For example, yeast cells cannot tolerate a concentration of alcohol much more than 12%. So beer and wine rarely contain more than 12% alcohol.

Top Tip
Don't forget, glycolysis is common to both aerobic and anaerobic respiration.

Quick Test

1. What is the most common respiratory substrate?

2. What is the first stage of respiration called?

3. Express the relative production of ATP in aerobic and anaerobic respiration as a simple whole-number ratio.

4. What are the end products of anaerobic respiration in animals and plants?

5. Why do we breathe heavily after strenuous exercise?

Answers 1. Glucose. **2.** Glycolysis. **3.** 19:1. **4.** Lactic acid in animals; ethanol and carbon dioxide in plants. **5.** To remove lactic acid from our body; to repay the oxygen debt.

Photosynthesis: two reactions

Key Facts
- Photosynthesis is the process by which plants make food using sunlight as a source of energy.
- There are two stages to the process: photolysis and carbon fixation.

Overview

During photosynthesis, plants take carbon dioxide from the air and water from the soil and combine them, using sunlight as a source of energy, to make glucose. There is a by-product, oxygen, which is released into the air.

Top Tip
Almost all life on Earth relies directly or indirectly on photosynthesis to trap energy.

oxygen is given out

light energy from the sun

carbon dioxide is taken in from the air

water is carried to the leaf from the soil

glucose is made in the leaves and taken to all parts of the plant

carbon dioxide + water	light energy	glucose + oxygen
raw materals	chlorophyll	*products*

Stage 1: Photolysis

Photosynthesis takes place in tiny cell structures called **chloroplasts**. They contain a complex green pigment called **chlorophyll** which uses the light energy to make ATP and to split water molecules. The process of splitting water is called **photolysis**, and it results in the formation of oxygen and hydrogen.

The oxygen diffuses out of the leaf, but the hydrogen is picked up by a carrier molecule and taken on to the second stage in the chloroplast.

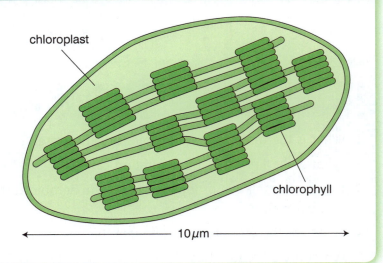

chloroplast

chlorophyll

10 μm

Stage 2: Carbon fixation

The hydrogen is carried to another part of the chloroplast where it combines with carbon dioxide, in a series of enzyme-catalysed reactions, to make glucose. This second stage is powered not by light, but by ATP, and is often referred to as the **carbon-fixation** stage. The glucose produced can be stored as starch, used to make cellulose for cell walls, or respired to produce more ATP for the cell. This is particularly important at night when the plants cannot photosynthesise. In fact, plants can go on to make thousands of other compounds from the simple products of photosynthesis, with the addition only of minerals from the water they absorb from the soil.

For example:

- magnesium is needed to make chlorophyll
- phosphorus is needed to make DNA and ATP
- iron is needed to make compounds involved in respiration.

Top Tip
Make sure you know all the details of both stages of photosynthesis.

Summary diagram

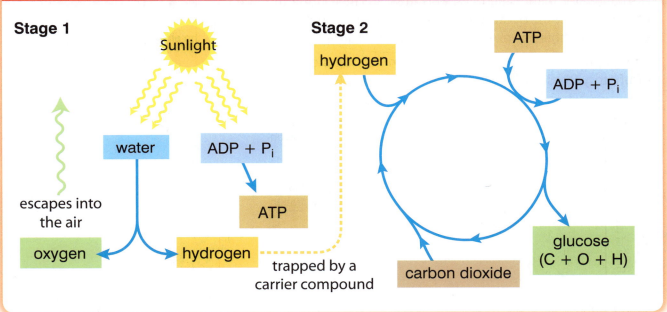

Quick Test

1. Where in the cell does photosynthesis take place?
2. What is the first stage of photosynthesis called?
3. For what purpose is the light energy used?
4. What two substances produced in the first stage are required for the second stage?
5. Why is the second stage called the carbon-fixation stage?

Answers **1.** Chloroplasts. **2.** Photolysis. **3.** To split water molecules and to provide the energy to make ATP. **4.** Hydrogen and ATP. **5.** Because carbon dioxide is taken from the air and combined chemically with other compounds.

25

Factors affecting photosynthesis

- The rate of photosynthesis is limited by only one environmental factor at any point in time.
- The factors that can affect the rate of photosynthesis are light intensity, temperature and the concentration of carbon dioxide.

Measuring the rate of photosynthesis

Rates of photosynthesis can measured in a variety of ways:
- By measuring the increase in dry mass of a plant over a period of time.
- By measuring the volume of oxygen given off over a period of time.
- By measuring the volume of carbon dioxide taken in over a period of time.

A small water plant called *Elodea* is ideal for measuring the rate of photosynthesis, because cut stems of the plant produce a stream of bubbles of oxygen which can be counted.

The graph below shows the results of an investigation into rates of photosynthesis using *Elodea*.

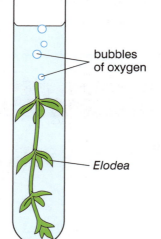

bubbles of oxygen

Elodea

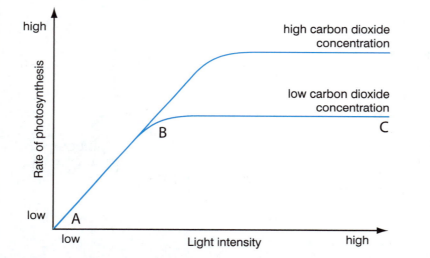

When there is no light the plant cannot photosynthesise. However, as light intensity increases, so does the rate of photosynthesis up to point B where some other factor is limiting the rate. To find out what that factor is the plant is given more carbon dioxide, or the temperature of the water is raised. One of these will cause a further increase in the rate of photosynthesis.

Top Tip
Make sure you understand the information given in the graph, because graphs like this often appear in exams.

Limiting factors

Plants need light, water and carbon dioxide to photosynthesise. Also, many of the chemical reactions in the second stage are catalysed by enzymes which are very sensitive to the temperature of their surroundings. Consequently, temperature, carbon dioxide concentration and light intensity all affect the rate of photosynthesis.

At any point in time, only one of these factors will limit the rate of photosynthesis.

Making the most of limiting factors

People who grow plants for a living, such as farmers, can make use of this information to make their plants grow faster and to allow them to grow plants earlier in the year.

By covering plants with polythene tunnels, or growing them in greenhouses, the temperature is raised and the plants are protected from damaging wind and pests.

If a paraffin stove is burned, this warms the plants and has the added benefit of providing them with extra carbon dioxide. Artificial lights provide extra light and warmth for plants.

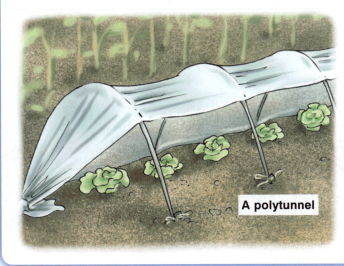

A polytunnel

Top Tip
Water does limit the rate of photosynthesis, but only indirectly. So, in an exam, don't give water as a factor which limits photosynthesis.

Quick Test

1. State three factors that can affect the rate of photosynthesis directly.
2. How many factors can limit the rate of photosynthesis at any point in time?
3. Suggest the major drawback of measuring the dry weight of a plant?
4. From the graph shown opposite, what factor is limiting photosynthesis between points
   a A and B
   b B and C?

Answers 1. Carbon dioxide concentration, light intensity and temperature. 2. One. 3. It is killed in the process. 4. **a** Light intensity; **b** CO_2 concentration

Ecosystems: food chains and food webs

Key Facts
- Food chains show the flow of energy and raw materials from one organism to another.
- Food webs are composed of a whole variety of interlinked food chains.

Ecosystems

The study of organisms in their natural habitats is called **ecology**. Ecologists study animals and plants living in **ecosystems**, and they are interested in how organisms are influenced by their **environment**. An ecosystem is the sum total of all the organisms and the places in which they live – their **habitats**. All the animals and plants in an ecosystem make up the **community**, and groups of single species make up individual **populations**.

Top Tip
Remember the formula: Habitat + Community = Ecosystem

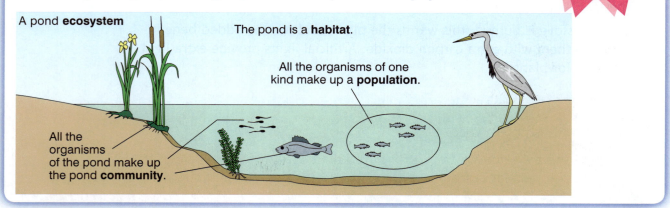

A pond **ecosystem**

The pond is a **habitat**.

All the organisms of one kind make up a **population**.

All the organisms of the pond make up the pond **community**.

Food chains

The Sun's energy trapped by plants in the process of photosynthesis is plundered by animals in a variety of ways; directly by animals that eat the plants and indirectly by animals that eat other animals.

This transfer of energy and food from one organism to the next can be shown as a **food chain**.

Plants are said to be **producers** because they make energy-rich food by photosynthesis. All animals and fungi are **consumers**, because they make use of this supply of energy.

Top Tip
Remember the meanings of all the words in bold. They are often asked in exams.

producer

primary consumer
herbivore
prey

secondary consumer
carnivore
predator

The animals at the beginning of a food chain eat plants, and they are called **herbivores**. Animals that eat herbivores are called **carnivores**. Further along the food chain there may be carnivores that eat other carnivores, or animals that eat both plants and animals. Humans belong to this last group, called **omnivores**.

Food webs

Most food chains are linked together in a complex way to form **food webs**. The position of an animal in a food web with respect to its eating habits and to its surroundings is called its **niche**.

An important group of organisms in a food web are the **decomposers** because they break down the dead remains of animals and plants and, as a result, play an important part in recycling. Fungi and bacteria are particularly important in this decomposition processes.

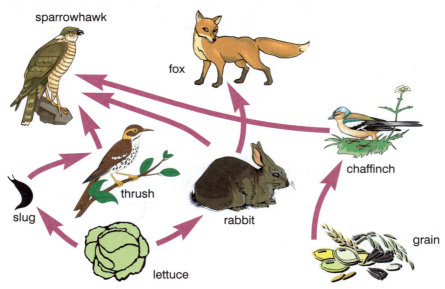

When organisms are removed from a food web the delicate balance is often disturbed and populations of other species may rise or fall as a consequence.

Quick Test

1. In the food web shown above name:
   a) a producer
   b) a herbivore
   c) a secondary consumer
   d) a predator and its prey.

2. What might happen to the population of rabbits if the foxes were removed?

3. In the food chain what do the red arrows represent?

Answers 1. a) Lettuce or grain. **b)** Slug, rabbit or chaffinch. **c)** Thrush or fox. **d)** e.g. Fox and rabbit, Sparrowhawk and chaffinch. **2.** It would increase. **3.** The transfer of energy, or food.

Ecosystems: energy loss

Key Fact

- As energy is passed from one organism to another along a food chain, much of it is lost.

The flow of energy

As the Sun's energy, in chemical form, is passed from one animal to the next along a food chain, each animal uses most of the energy for its own purposes. So, by the time it gets eaten by another animal, very little of the energy is left to pass on.

Consequently, food chains are rather short, and there are very few animals at the top end. These are the **secondary** or **tertiary consumers** such as hawks, foxes, cats and sharks. They are not left with much of the original energy trapped by the plants because it has been lost as heat or waste products such as urine and faeces. This is why tertiary consumers have such small populations.

Decomposers trap the energy lost from the food chain, but they don't always succeed. When we burn coal, oil and gas, we are releasing the Sun's energy trapped by plants and animals millions of years ago which has not been completely decomposed.

1. The caterpillar eats the leaf.

2. Some of the energy in the leaf is lost in waste which passes through the digestive system of the caterpillar.

3. Some of the energy is lost as heat as the caterpillar moves about.

4. Only a small proportion of the energy is stored in the body of the caterpillar. This is the only energy available to the animal that eats it.

Pyramids of numbers

This flow of ever-decreasing quantities of energy along a food chain usually results in a decrease in the numbers of organisms. The populations of each organism in a food chain can be represented by a **pyramid of numbers**, where the numbers of organisms at each link in the food chain decrease.

Pyramids of numbers are not always true pyramids in shape.

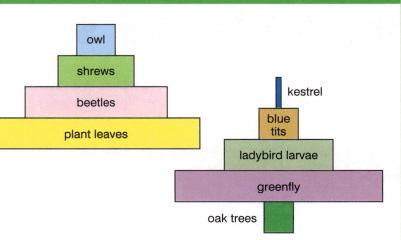

Pyramids of biomass

If all the organisms at each link in a food chain are weighed, no matter what size they are or how many of them there are, a true pyramid is almost always obtained. This is called a **pyramid of biomass**.

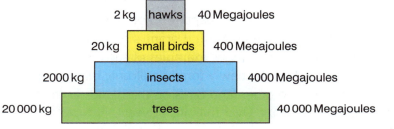

Pyramids of energy

The most reliable form of pyramid is one that represents the total energy content of organisms at each link in the food chain. As energy is always lost at each stage, so the shape of a **pyramid of energy** is always maintained.

Top Tip
Don't forget the different ways in which energy is lost from food chains.

Quick Test

1. Which of the pyramids of numbers shown is the best representation of a food chain that starts with an oak tree and ends with a hawk?

2 Why are there rarely more than four or five links in a food chain?

3 Describe two ways in which energy is lost from a food chain.

Ecosystems: variety of species

Key Facts
- Biodiversity is the variety of organisms in an ecosystem.
- Stable ecosystems usually have high biodiversity.

Species

Organisms are said to belong to the same species if they can breed with one another and produce fertile offspring. There is a huge variety of species in most ecosystems, and a wide variety of individuals within each species. Ecologists refer to the variety of species and individuals within a species as **biodiversity**.

Biodiversity

Top Tip
Darwin's finches are mentioned in the exam syllabus, so make sure you know about them.

Biodiversity results from the adaptations of animals and plants to every niche available to them on Earth. Darwin's finches, which live on the Galapagos Islands, are famous examples of such biodiversity. Their ancestors arrived on the newly formed volcanic island around 5 million years ago. There were no other land birds on the islands at the time, so they had no other **competitors**. Over millions of years, **natural selection** modified their bodies, adapting them to the wide range of ecological niches available. This was **evolution** at work, and Charles Darwin, who visited the islands, was the first person to recognise and describe this process.

Now, from the one original ancestral species, there are 13 different species of finches on the islands, all slightly different from one other; some eating seeds, some eating fruits or insects, and some even drinking blood. Blood is an important source of water on one of the smaller islands which has no fresh water. There, plants too are adapted to the dry conditions, where thick waxy cuticles, small leaves and extensive roots all help save water.

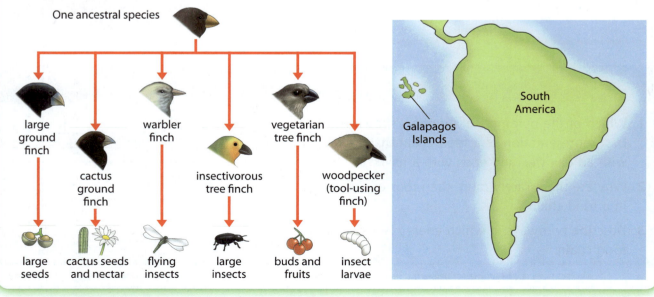

Habitat destruction

Biodiversity on Earth is in severe decline because of the activities of humans. Even on the remote Galapagos Islands, some of the animals and plants face extinction because of human influence. For example, rats, introduced by humans, eat vulnerable birds' eggs, and goats graze on vegetation, removing fragile plant species.

The huge forests of neighbouring South America are being cut down at a remarkable rate. This reduces biodiversity and has many other impacts on Earth, including climatic change. The human species is making this planet an increasingly difficult place in which to live, both for itself and for many other animals and plants.

Pollution of the air, soil and water also has a major impact on biodiversity. There has never been a time in the history of the Earth when so many species have become extinct in such a short period of time, and there seems no end in sight to this destruction.

Quick Test

1. Give an example of a human activity that could reduce biodiversity.

2. A horse bred with a donkey produces a mule which is always infertile. What does this tell you about horses and donkeys in terms of species?

3. Describe two adaptations of Darwin's finches.

Answers 1. Deforestation, mining, overfishing, road-building. 2. They belong to different species. 3. Large strong beaks for breaking nuts; small sharp beaks for picking up insects.

Ecosystems: animal behaviour

Key Facts
- In any ecosystem, animals compete for food, water, mates, territory and shelter.
- Animal behaviour is adapted to help ensure survival.

Competition

Because there is only a limited supply of food, mates and places to live, animals compete with one another. Even members of the same species will compete. Rams will fight with one another to gain the prize of a ewe with which to mate; lions will fight over who gets the tastiest morsel of meat. Life is a constant battle in which only the fittest and best-adapted will survive.

Evolution ensures that the best adapted animals and plants go on to reproduce and bring new, well-adapted offspring into the world.

Top Tip
Don't forget that plants compete as well, for light, water, minerals and space to grow.

Choice chambers

An example of simple animal behaviour is shown by woodlice. (Woodlice are often called slaters in Scotland.)

Woodlice can be found in the garden and brought into the lab to study. They are crustaceans, and are related to crabs and lobsters.

They like to live in damp, dark and moist places, and this can be tested using a simple piece of apparatus called a **choice chamber**.

A chamber is set up in which one environmental factor is changed. Everything else must be kept constant. We call all the factors which we control **variables**. So, if we are measuring the effect of different light intensities on woodlice, one side might be dark and the other light. But all the other variables would be kept the same, e.g. the species of woodlouse used, the temperature, the presence or absence of food and the moisture content of the air (**humidity**). A number of woodlice are then placed in the chamber and left for 1 hour. Later, you check to see where the woodlice have congregated.

Top Tip
The response of woodlice to light and humidity is mentioned in the exam syllabus.

Reliability and validity

It is important that more than one woodlouse is used and that the experiment is repeated. This ensures that the results are **reliable**.

If the humidity and the light were changed in each chamber then the investigation would not be **valid** because you would be unable to separate the influences of light and humidity on the behaviour of the woodlice. If the woodlice ended up in one chamber which was moist and dark, you would not know whether it was the moisture or the darkness, or both, which had caused the woodlice to move.

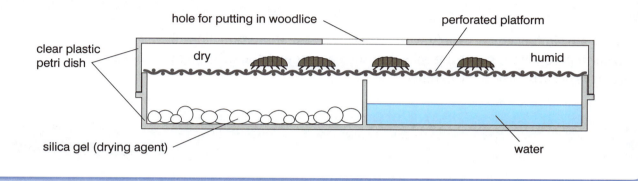

Survival behaviour

Woodlice survive better in damp, cool and dark conditions. There, they avoid predators and do not lose water quickly. There's also likely to be more food around. How do they know where to go? The answer is that they don't. All they do is move about more rapidly and randomly when conditions are unsuitable. When conditions are suitable, that type of behaviour is switched off and the woodlice sit tight or move only very slowly. Because of that simple mechanism, woodlice tend to find, and stay in, the environment that suits them best.

Quick Test

A choice chamber is set up with one side moist and the other side dry.
Ten woodlice are placed in the choice chamber and left for 1 hour undisturbed.

1. Where should the woodlice be placed?

2. Why were ten woodlice used rather than one?

3. State two factors that should be kept constant during the investigation.

4. Where would you expect to find the woodlice at the end of the investigation?

5. What advantage is it to the woodlice to behave in this way?

Answers 1. In the middle or equal numbers on each side. 2. To improve reliability. 3. Temperature and light intensity. 4. On the moist side. 5. They reduce water loss from their bodies.

Fertilisation

- Sex cells are called gametes.
- In mammals, sperm are produced in the testes and eggs in the ovaries.
- In flowering plants, pollen contains the male gamete and ovules contain the female gamete.
- Fertilisation is the fusion of the nuclei of the gametes.
- A fertilised egg is called a zygote.
- Sexual reproduction brings about variation in species.

Sexual reproduction

One major influence on variety within species is sexual reproduction. During sexual reproduction, genes are mixed in such a way that the offspring produced are never exactly like their parents.

The human reproductive system

The penis delivers sperm to the vagina. The sperm swim up to the oviduct where they fertilise the egg. The egg starts to divide and is carried down to the uterus where it settles in the wall. The baby grows for nine months before it is delivered to the outside world.

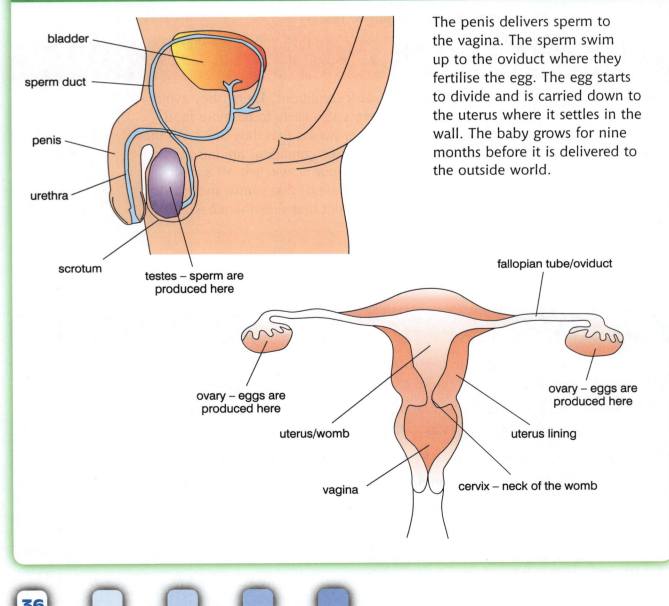

bladder

sperm duct

penis

urethra

scrotum

testes – sperm are produced here

fallopian tube/oviduct

ovary – eggs are produced here

ovary – eggs are produced here

uterus/womb

uterus lining

vagina

cervix – neck of the womb

The reproductive system of a flowering plant

In flowering plants the male and female parts are often found in the same flower. You might imagine that this is to ensure easy self-fertilisation, but this is not the case. Plants are adapted to ensure they do not self-fertilise, because self fertilisation reduces variety.

An insect pollinated flower

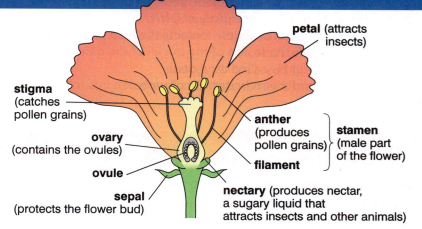

petal (attracts insects)

stigma (catches pollen grains)

ovary (contains the ovules)

ovule

sepal (protects the flower bud)

anther (produces pollen grains)

filament

stamen (male part of the flower)

nectary (produces nectar, a sugary liquid that attracts insects and other animals)

Gametes

Gametes (sex cells) are very special cells in two respects:

- They contain only half the genetic material of all other cells in the body.
- They are genetically different from one another and all other cells in the body.

Body cells contain two complete sets of genes. This is because they arise from the fusion of two gametes, each of which has a single set of genes. This means that we have two genes for almost every characteristic. Usually, these genes are identical, but sometimes they are not. You will learn more about this in the next few pages.

Top Tip
The most important thing about sexual reproduction is that it results in variety of offspring.

Fertilisation

Fertilisation is the fusion of the nuclei of two gametes. Since every gamete has a unique genetic makeup, and since fertilisation is random, organisms produced by sexual reproduction are never the same, the only exception being identical twins.

male gamete + female gamete = **zygote** (a fertilised egg)
(one set of genes)   (one set of genes)   (two sets of genes)

Quick Test

1. Where are sperm and eggs produced in mammals?
2. What structures carry the male gametes in flowering plants?
3. What is a zygote?
4. Why do body cells contain two sets of genes?

Answers 1. In the testes and ovaries. 2. Pollen grains. 3. A fertilised egg. 4. Because they arise from the fusion of two gametes, each of which has a single set.

Chromosomes and DNA

Key Facts

- Chromosomes contain genetic information that gives rise to an organism's characteristics.
- Chromosomes consist mostly of DNA.
- DNA has millions of simple molecules called bases which, when linked together as a chain, make up the genetic code.
- A sequence of bases makes a gene.
- Genes control the manufacture of protein.

Chromosomes

Genes are found on thread-like structures called **chromosomes**, and chromosomes are found in the nuclei of all cells with very few exceptions. Each species of organism has a fixed number of chromosomes which may vary from just a few to over 100.

Organism	Chromosome number
Mouse	40
Pea plant	14
Kangaroo	12

This chromosome number is always an even number, because chromosomes exist in pairs. Humans have 46 chromosomes in each body cell (23 pairs), and human chromosomes contain around 30 000 genes in total. Every time a cell divides in our body, the chromosomes are copied perfectly and the copies passed on to the new cells.

Sex determination

Two chromosomes are particularly interesting, because they decide the sex of an organism. These are called the sex chromosomes, and there are two types:
a **Y chromosome** and
an **X chromosome**.
The Y chromosome is shorter than the X chromosome.

In humans if you have two X chromosomes you are a woman, and if you have an X and a Y chromosome you are a man. Many other species of organism have a similar arrangement. Males produce two kinds of sperm: X and Y. Females only produce X eggs. So if a Y sperm fertilises an egg a male child results. If an X sperm fertilises an egg a female child results.

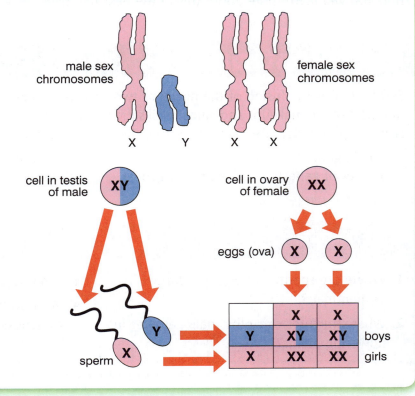

DNA and genes

Chromosomes are packed with genes. Genes are made of **DNA**, which stands for **deoxyribonucleic acid**. The DNA contains the genetic code to build an organism and is unique to every individual.

DNA is made up of many millions of molecular units called **bases**. There are four different types of base, and a gene is a sequence of these bases. They dictate the type of **protein** made by a cell.

The coloured bars represent the four different kinds of base

Top Tip
DNA is found in almost every organism on Earth and the coding system is universal.

Proteins

Proteins are the basic building blocks of all living things. Enzymes are proteins, so all the metabolic reactions taking place in cells are controlled by proteins. Most hormones are proteins, and hormones are vital in the control of behaviour, growth and development.

Some examples of other important animal proteins include:
- **antibodies** that protect us from disease
- **actin** and **myosin** fibres of muscles
- **haemoglobin**, which carries oxygen round the body
- **keratin**, which is the main component of nails and hair
- **collagen**, which gives strength to our skin, tendons, ligaments and bones.

Top Tip
The sequence of bases in a strand of DNA dictates what proteins are made by the cell.

Quick Test

1. Where are chromosomes found?
2. What are chromosomes made of?
3. What is a gene?
4. How many different types of bases are there?
5. Name three important substances that are made of protein.

Answers 1. In the nucleus of a cell. 2. DNA. 3. A strand of DNA or a sequence of bases that code for a protein. 4. Four. 5. Enzymes, hormones and antibodies.

Division of the nucleus

- Body cells all contain two sets of chromosomes.
- When cells divide, their genetic material must be copied and passed on to the new cells.
- When body cells divide to form gametes, by a process called meiosis, each gamete gets only one set of chromosomes.
- During meiosis the genes are also shuffled so each new gamete is different from the rest.

Meiosis

When gametes are formed they are all different from one another and have only half the normal number of chromosomes. The process of shuffling genes and sharing out chromosomes to gametes is called meiosis. Meiosis is accompanied by two cell divisions, so that after it is completed, four sex cells are produced each having half the number of chromosomes of the parent. When the gametes fuse at fertilisation, the normal chromosome number is regained.

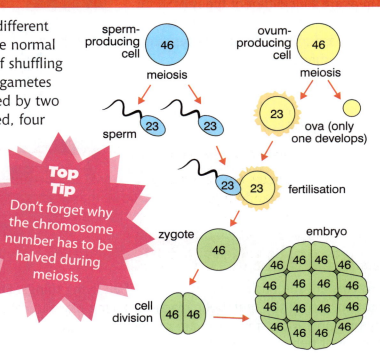

Top Tip

Don't forget why the chromosome number has to be halved during meiosis.

Table to show sites of meiosis		
	Male	**Female**
Mammals	Testes	Ovaries
Flowering plants	Anthers	Ovaries

Sharing out chromosomes

Most animals and plants have two sets of matching chromosomes. This is simply because they obtained one set from their male parent and one set from their female parent.

During meiosis all chromosomes find their opposite partner and line up along the equator of the cell. When they line up along the equator of the cell, the chromosomes can do so in a variety of ways.

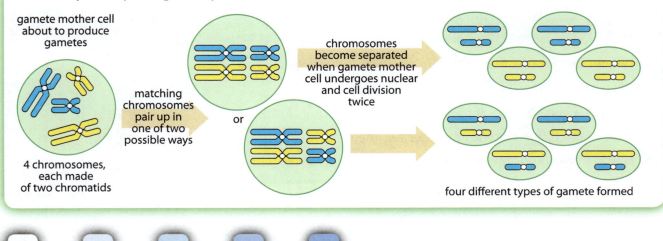

gamete mother cell about to produce gametes

matching chromosomes pair up in one of two possible ways

or

chromosomes become separated when gamete mother cell undergoes nuclear and cell division twice

4 chromosomes, each made of two chromatids

four different types of gamete formed

Variety of gametes

The table shows the relationship between chromosome number and the number of different ways in which they can line up and the number of different gametes that can be produced.

In humans, with 23 pairs of chromosomes, there are over 4 million different ways in which they can line up along the equator of the cell. So this results in considerable mixing of the

Number of chromosome pairs	Number of ways in which they can line up	Number of different gametes which result
2	2	4
3	4	8
4	8	16
5	16	32
6	32	64

genetic information. Once aligned along the equator, they are pulled apart and go to opposite ends of the cell. The cell then divides, so that each new cell has only half the original number of chromosomes, selected at random from each pair.

Variety of life

Because all gametes are genetically different, and because fertilisation is a random process, it follows that all offspring vary in their appearance. This is why we all look slightly different from our parents and why there is no one else like us on Earth, with the exception of identical twins. This variation is the key to the process of evolution.

When the environment changes, all the members of any one species risk becoming extinct, unless some of them are better adapted than others. For example, it was thought that the viral disease myxomatosis would wipe out the entire UK rabbit population in the 1950s. But it didn't, because rabbits varied and some were resistant to this killler disease.

Top Tip
Remember, the random assortment of chromosomes during meiosis leads to variation in offspring.

Quick Test

1. How many sets of chromosomes are found in gametes?

2. How many sets of chromosomes are found in body cells?

3. Name the process that results in the formation of gametes.

4. What happens to the genes during this process?

5. What happens to the chromosome number during this process?

6. How many different gametes could you get from an organism with a chromosome number of 16?

Answers 1. 1. 2. 2. 3. Meiosis. 4. They are shuffled. 5. It is halved. 6. 256 (8 pairs of chromosomes line up 128 different ways)

Genes and alleles

- Sexually reproducing organisms have pairs of genes controlling each characteristic, one from each parent.
- These pairs of alternative genes are called alleles.

Genetics

All the features of an organism, even the basis of its behaviour, are controlled by genes. Sexually reproducing organisms have pairs of genes for each characteristic because they inherit one from each parent. Most often these genes are identical, because they code for a protein that is essential for life. However, with some characteristics, e.g. colour of hair, there can be a variety of controlling genes. These alternative genes are called **alleles** and the study of their inheritance is the basis of **genetics**.

Gregor Mendel

The first person to carry out successful genetics experiments, around 150 years ago, was an Austrian monk called Gregor Mendel. He took true-breeding tall pea plants and crossed them with true-breeding dwarf plants, and found that all the offspring from this first generation, called the **F$_1$ generation**, all grew tall. The dwarf characteristic had mysteriously disappeared. When he allowed the new tall plants to cross with one another, some of the next generation (**F$_2$ generation**) of plants were dwarf plants. The dwarf characteristic had reappeared!

Being a scientist, as well as a monk, Mendel counted the different types and discovered there was a ratio of approximately 3 tall to 1 dwarf in this second generation of peas. He went on to lay the foundations of modern genetics, but, sadly, his published paper was ignored by the scientific community for many years and he died in obscurity.

Mendel's crosses can be explained as follows:

The tall plants were **true-breeding**. This means they had inherited a gene for tallness from each parent. The dwarf plants had inherited a gene for dwarf height from each parent. When they were crossed, their offspring must have obtained a tall gene from one parent and a dwarf gene from the other parent. This generation of tall plants ignored the instruction from the dwarf gene and grew tall. Because of this we say that the tall gene is **dominant** and the dwarf gene is **recessive**.

Top Tip
Make sure you can explain why some inherited features seem to skip generations.

Top Tip
Make sure you understand the difference between alleles and genes.

Mendel's pea plant cross

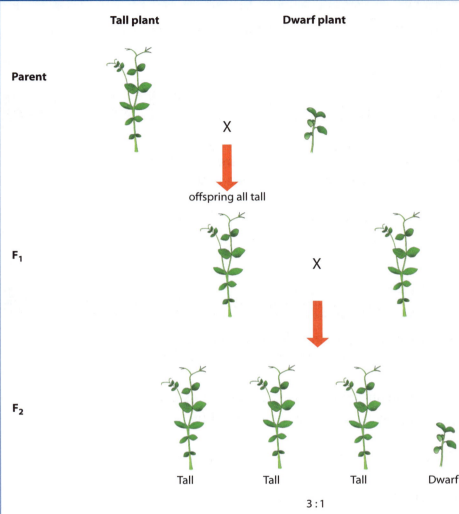

Tall plant   Dwarf plant

Parent

X

offspring all tall

F₁

X

F₂

Tall   Tall   Tall   Dwarf

3 : 1

Letters for genes

Alternatives of genes (**alleles**) are usually represented by letters of the alphabet.

So **T** for tall and **t** for dwarf are each alleles for height. We use the same letters of the alphabet for a particular characteristic, and the capital letter denotes the dominant gene; while the small letter denotes the recessive gene.

Quick Test

1. What term is used to describe the alternative of a gene?

2. Give an example of two alleles.

3. What does a capital letter for an allele indicate about that allele?

4. What does a small letter for an allele indicate about that allele?

Answers 1. An allele. 2. T and t for height in peas. 3. It is dominant. 4. It is recessive.

Monohybrid crosses

> **Key Fact**
> * Parents in experimental monohybrid crosses are often true breeding (homozygous) and show different phenotypes.

Genetic crosses

Since Mendel's first experiments with pea plants, geneticists have carried out millions of crosses between organisms to determine how genes are passed from one generation to the next. When many genes are studied at one time, this becomes a very complex process. At Intermediate 2 level, you need to be able to follow the inheritance of only one feature at a time, and this is described as a **monohybrid cross.**

Genetic terminology

To make the description of crosses between organisms clearly understood, a number of different terms are used. You have met some of these terms already. Here are some more:

* **Phenotype**: The appearance of an organism; often with respect to its genes.
* **Genotype**: The genes an organism possesses; usually represented by letters of the alphabet.
* **Heterozygous**: Having a pair of alleles that are different from one another eg Tt.
* **Homozygous**: Having a pair of alleles that are identical, e.g. TT or tt.

Crossing mice

All the questions on the next page relate to this cross between two different mice. A true-breeding brown mouse is crossed with a true-breeding white mouse and all the offspring are found to be brown.

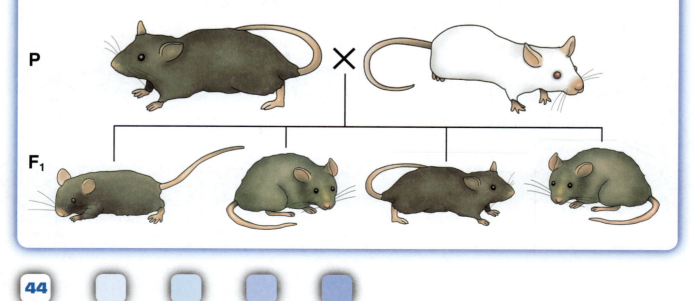

Quick Test

1. What are the phenotypes of the parents (the colours)?

2. Which characteristic is dominant? How are you able to tell?

3. What are the genotypes of the parents? (use **B** for brown and **b** for white colouration)

4. What are the genotypes of the gametes produced by each parent? (one letter for each gamete)

5. What is the genotype of the F_1 generation? (two letters)

6. What is the phenotype of the F_1 generation? (colour)

7. What is the genotype of a true breeding brown mouse?

8. Do you think a mouse from the F_1 generation would be true-breeding?

9. What colour is a heterozygous mouse?

10. State the genotype of a white mouse. Is it heterozygous?

Two heterozygous mice from the F_1 generation are crossed

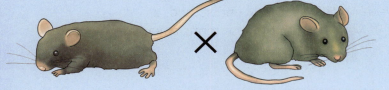

11. What are the possible genotypes of the gametes produced?

These gametes can fertilise one another in four different ways:

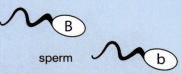

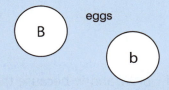

eggs

sperm

12. Write down the four different ways in which these gametes can combine.

There is a neat way of working out the ways in which the gametes can combine using a **Punnett** square:

Gametes	B	b
B	BB	
b		

You simply complete the four boxes in the square to give the F_2 generation. One is done for you.

13. From the Punnett square, state the four genotypes of the F_2 generation.

14. What are the phenotypes of each of these mice?

15. How many white and brown mice would you expect in the F_2 generation?

Answers 1. Brown and white. 2. Brown, because all the F_1 generation are brown and not white. 3. BB and bb. 4. B and b. 5. Bb. 6. Brown. 7. BB. 8. No. 9. Brown. 10. bb No 11. B or b from each parent. 12. B with b, B with B or b with b. 13. BB, Bb, Bb and bb. 14. Brown, brown, brown and white. 15. 3 brown to 1 white.

More about inheritance

Key Facts

- When two different alleles both affect the phenotype of an organism they are said to be co-dominant.
- When many genes affect one characteristic of an organism, that feature is said to be polygenic.

Co-dominance

In some cases, one allele is not dominant over another. Instead, both have an equal effect on the phenotype of an individual. In this case we say they are **co-dominant**. For example, in snapdragons, if red flowered plants are crossed with white flowered plants the F_1 flowers are all pink in appearance, because both the red and white alleles affect the phenotype.

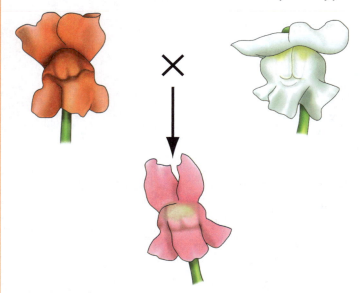

The alleles cannot be called R and r or W and w because that would imply that one was dominant to the other. Instead, we use capital letters for both co-dominant alleles.

So the allele for red is R and the allele for white is W.

Red flowers are therefore RR.

Top Tip
Always remember to use two different capital letters to denote two co-dominant alleles.

Quick Test

1. What are the genotypes of white flowers and pink flowers?

2. If pink flowers were crossed with one another, what proportions of colours would you expect in the F_2 generation? Use a Punnett square to help you.

3. What offspring, and in what proportions, would you expect from a cross between a pink flowered plant and a red flowered plant?

Answers 1. WW and RW. 2. One white, one red and two pink. 3. Equal numbers of pink and red flowered plants.

Polygenic inheritance

Top Tip
Features like height and weight in humans are polygenic.

Many characteristics of animals and plants, such as height and skin colour in humans, are controlled by more than one pair of alleles. These characteristics are said to be **polygenic.**

Quick Test

Imagine that colour of skin was influenced by two pairs of alleles on two separate chromosomes, and that each allele had a different effect on the phenotype, i.e. the alleles were not completely dominant over one another.

Allele **A** gives 8 units of colour   Allele **a** gives 2 units of colour

Allele **B** gives 5 units of colour   Allele **b** gives 1 unit of colour.

1. Write down all the possible combinations of **A**, **B**, **a** and **b** alleles, given that there must always be two **A**s and two **B**s. You cannot have three **A**s and one **B**, or four **B**s and no **A**s. The first few are completed for you.

Genotype	Phenotype (units of colour)
AA BB	26
AA Bb	22
AA bb	
Aa BB	

2. From your table, how many different shades of colour are possible when there are two pairs of alleles controlling colour?

3. If there were three pairs of alleles, how many different shades of colour might be possible?

Answers
1.
Genotype	Phenotype (units of colour)
AA BB	26
AA Bb	22
AA bb	18
Aa BB	20
Aa Bb	16
Aa bb	12
aa BB	14
aa Bb	10
aa bb	6

2. 9 3. Up to 27 different shades

Natural selection

Key Facts
- The best adapted organisms are more likely to survive and pass their useful characteristics on to the next generation.
- Over long periods of time, natural selection leads to the gradual evolution of new species.

Evolution

Humans have been selecting animals and plants for breeding for thousands of years, but nature has been doing the same thing for billions of years. We call this **natural selection** and the process was first described by Charles Darwin.

Darwin noticed the differences in breeds of domesticated pigeons and dogs and realised they had been brought about through selection by their breeders over many generations. He also noticed that wild animals and plants varied and concluded these variations must have happened after the beginning of life on Earth. He saw evidence of this on the Galapagos Islands, where it was clear one species of finch had arrived millions of years ago, but had then gone on to evolve into 13 different species, occupying many different niches which were available on the uninhabited islands. Many other animals on the islands show similar changes. (See page 32.)

Up to that time, most people thought that species were unchanging, but Darwin thought otherwise and proposed a mechanism by which gradual change, or evolution, might occur. His theory can be summarised as a number of observations and conclusions.

- Observation 1: Populations of organisms have the potential to increase in size because two parents can always produce many more than two offspring during their lives.
- Observation 2: However, most populations remain relatively stable over long periods of time.
- Conclusion 1: Many offspring must die before being able to reproduce.
- Observation 3: Offspring of any sexually reproducing organisms are different from their parents and from one another.
- Conclusion 2: The offspring with advantageous characteristics are more likely to survive and pass these characteristics on to their descendants.
- Conclusion 3: Over long periods of time, the accumulation of a series of small differences can result in the formation of new species.

The Peppered Moth

We now have overwhelming evidence that Darwin was correct. One of the many examples of evolution in action is the changes that have occurred to the Peppered Moth in the last 200 years.

The moth has a speckled grey coloration that camouflages it against lichen-covered tree trunks. However, black moths turn up from time to time as a result of mutations. Mutations are unplanned genetic changes that are usually harmful. The black moths are easily spotted by predators and eaten.

In the nineteenth century, the Industrial Revolution caused the blackening of tree trunks with soot from chimneys. Consequently, the black moth became better camouflaged than the peppered one. After a period of time, the black moth was common and the peppered variety uncommon. In other words, a tiny step in evolution had taken place: an organism had changed colour in response to a change in its environment. Humans too have changed skin colour over thousands of years, to better adapt to local conditions.

Top Tip
Take time to remember the Peppered Moth story as it is specifically mentioned in the exam syllabus.

dark form enjoys selective advantage on soot-covered trunk in polluted area

light form enjoys selective advantage on lichen-covered trunk in non-polluted area

Quick Test

1. Who first proposed the theory of natural selection?
2. What birds on the Galapagos Islands provided Darwin with evidence of evolution?
3. What colour is the Peppered Moth in unpolluted areas?
4. What colour was the Peppered Moth near cities during the Industrial Revolution?
5. How might you check whether these two varieties belonged to the same species?

Answers 1. Charles Darwin. 2. Finches. 3. Speckled grey. 4. Black. 5. Breed them and check whether any offspring are fertile.

Selective breeding

Key Facts
- Humans have bred domestic animals and crop plants for thousands of years to obtain desirable characteristics.
- The effects of the environment on phenotype are not inherited.
- There are two categories of variation: continuous and discontinuous.

Picking the best

Humans have bred organisms for thousands of years, without understanding the principles of genetics that Mendel discovered. They did this because they wanted the best organisms to work for them, or the best organisms to eat. For example, cows have been bred from wild cattle to provide high volumes of high quality milk, or to provide quick growth and lean meat. Wheat has been bred to provide high yield, quick growth, resistance to disease and to drought.

This process is called **selective breeding** and it is very simple. A male and female are chosen, each with desirable characteristics. They are crossed, and when the young are born, those not wanted are discarded and those with the desired characteristics are kept for further breeding. Over many generations, major changes can occur.

Top Tip
Selective breeding takes a long time and results are never guaranteed.

The effect of the environment on phenotype

It was thought for many years that changes to animals and plants, brought about during their lives, would then be inherited. But this is not the case.

For example, if you train a dog to catch small animals and bring them back to you, this characteristic will not be passed on to its offspring. If you feed a plant with fertiliser so that it grows better, this feature will not be passed on to its offspring. The reason for this is simple: training dogs and giving plants fertiliser so they grow better has no effect on their genes. Therefore characteristics acquired during the lives of animals and plants are not passed on to the next generation.

The appearance of an organism and the way it behaves, i.e. its phenotype, results from the combined effect of inheritance and environment, sometimes called 'nature and nurture'.

Top Tip
Don't forget the importance of 'nature and nurture'.

Continuous and discontinuous variation

Differences in phenotype can be divided into two broad categories: continuous and discontinuous.

Continuous variation results from polygenic inheritance where many different phenotypes are possible across a whole range of sizes and shapes, e.g. heights or weights of organisms, or skin colour in humans.

Discontinuous variation results from monohybrid inheritance in which only a few very distinct features are possible, e.g. height in pea plants has only two categories: tall or dwarf; blood groups in humans have only four different categories: A, B, AB or O.

Bar graphs and histograms

It is important, when you collect data, that you can display the data in the form of a graph for ease of understanding. Two types of graph, often used and often confused, are bar graphs and histograms.

The table below summarises the use of these.

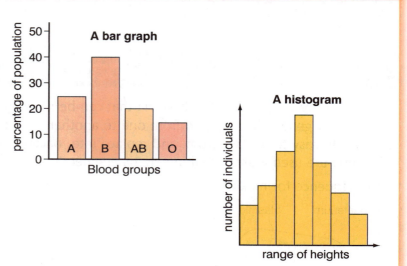

Type of graph	Type of data
Bar graph	Discontinuous
Histogram	Continuous

Quick Test

The graph shows the distribution of the resting heart rate in a human population.

1. Is the graph a histogram or bar graph?

2. What is the range of resting heart rates shown in the graph?

3. What percentage of the population has a heart rate of less than 50 beats per minute?

4. If a person has a heart rate of 75bpm, what is the length of one heart beat?

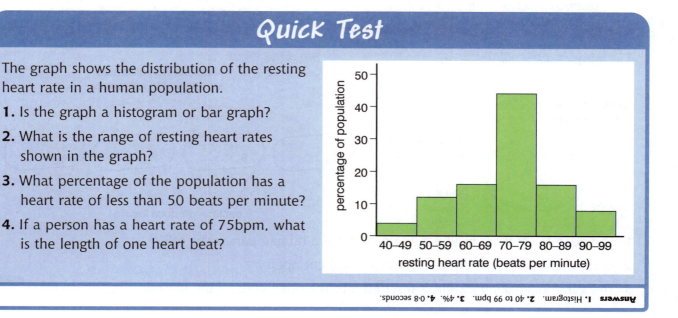

Answers 1. Histogram. 2. 40 to 99 bpm. 3. 4%. 4. 0·8 seconds.

Genetic engineering

- Genetic engineering involves the transfer of genes from one species of organism to another to get the second organism to produce a specific protein, e.g. an enzyme or hormone.
- Bacteria are often used because they have rings of DNA called plasmids, which are relatively easy to remove, alter and replace.

The Universal Code

The genetic code is read in the same way by every living organism on Earth. For this reason, we can now transfer parts of the code from one organism to another to reprogram the organism that has received the new code.

For example, human **insulin** and **human growth hormone** are both now produced by bacteria that have been **genetically engineered**.

The process involves enzymes that are used to open cells, cut DNA strands in specific places and insert new pieces of DNA.

Top Tip
Make sure you know what a plasmid is and how it is used in genetic engineering.

Plasmids: gene carriers

Plasmids are rings of DNA in bacteria that can be used by the bacteria to transfer genetic information from one to another. So plasmids are relatively easy to remove from, and replace into, bacterial cells, and bacteria are used a great deal in genetic engineering.

The sequence for making genetically engineered insulin is as follows:

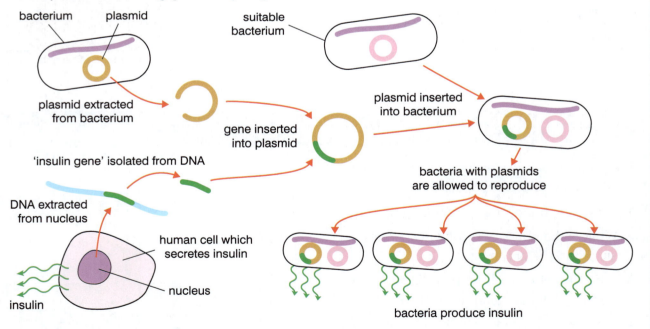

1. The gene for making insulin is removed from a human pancreatic cell.
2. A plasmid from a bacterium is removed and cut open.
3. The gene for making insulin is inserted into the open ends of the plasmid.

4 The plasmid is sealed and returned to the bacterium.

5 The bacterium divides many times to produce millions of genetically identical bacteria.

6 The bacteria are grown in a fermenter and produce a large volume of insulin which can then be extracted and purified.

This sounds much easier than it is. In fact, the success rate of this process is less than 1 in a million.

Genetic engineering: for and against

There are heated arguments about genetic engineering, with strong views expressed on both sides.

Those in favour claim that it is quick and very precise, i.e. only genes that are required are transferred from one organism to another, and these genes act in a precise way. So, for example, if the gene for the manufacture of human insulin is transferred from a human cell to a bacterial cell, the bacterium will make human insulin and not its own version of insulin. Another enormous advantage illustrated by this example is that, unlike in selective breeding, there are no species barriers to this process; genes from humans can be transferred to bacteria, and genes from fish can be transferred to strawberries!

Those who argue against genetic engineering do so from a variety of positions. Religious people claim we should not be tinkering with God's creation. Others argue that genetically engineered organisms may pass their altered genes to wild species and bring about unpredictable changes. Also, can we be certain that genetically engineered plants and animals are safe to eat?

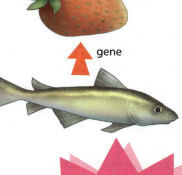

gene

Top Tip
The syllabus says that you should know about the advantages and disadvantages of genetic engineering.

Comparison of selective breeding and genetic engineering

	Selective breeding	Genetic engineering
Speed	Slow	Quick
Precision	Low	High
Species specificity	Within one species only	Across different species

Quick Test

1. Name two human hormones produced by genetically engineered bacteria.

2. What is a plasmid?

3. Where is insulin made in the human body?

4. What kind of substance is used to cut and glue DNA strands?

Answers 1. Insulin and human growth hormone. 2. A ring of DNA found in bacteria. 3. The pancreas. 4. Enzymes.

Food

- The main food groups are carbohydrates, proteins, fats, vitamins and minerals.
- Carbohydrates are composed of sugar molecules.
- Proteins are composed of amino acids.
- Fats are composed of fatty acids and glycerol.

The need for food

All organisms need food as a source of building materials and energy. Plants make their own food using the simple molecules carbon dioxide and water, and minerals from the soil. Animals and fungi cannot photosynthesise, and have to obtain their food from other organisms.

Carbohydrates and fats

Carbohydrates and fats are required mostly as a source of energy. Fats are particularly rich in energy, providing roughly two times more energy than carbohydrates. However, they have other functions too; fats are the main component of all cell membranes, and in mammals, some hormones are fatty substances. Also, fatty layers act as insulation in many mammals, e.g. whales.

Proteins

Proteins can also be used for energy, and have approximately the same energy content as carbohydrates. We get energy from the breakdown of excess proteins, and if we are starving we will start to break down body protein. However, proteins are much more important as the building materials for bodies (see page 39).

Vitamins and minerals

These are required in tiny quantities for a whole range of functions in the body. In particular, they are needed to assist in many enzyme-catalysed reactions. If we become short of any particular vitamin or mineral then the functioning of the body will deteriorate:

- A lack of iron will result in anaemia (lack of red blood cells).
- A lack of calcium can lead to problems with teeth and bones.
- A lack of vitamin C can lead to scurvy.

Vegetables and fruit are particularly rich in vitamins and minerals and as a consequence we are recommended to include plenty in our daily diet.

Composition of foods that contain energy

	Carbohydrates	Fats	Proteins
Chemical elements present	carbon (C) oxygen (O) hydrogen (H)	C, O and H	C, O, H and nitrogen (N)
Basic units from which they are built	sugar molecules	fatty acids and glycerol	amino acids
Molecular structure	glucose molecules in a chain to form e.g. starch or cellulose 	glycerol fatty acids	amino acids in a chain to form a protein
Dietary sources	Bread and potatoes	Butter and cheese	Meat and eggs

Top Tip
The information in the table is well worth memorising.

Quick Test

1. Which food groups can be used as a source of energy?
2. If you eat equal weights of fats, carbohydrates and proteins, which will provide the most energy?
3. Name two important substances that are made of protein.
4. Name the four chemical elements common to all proteins.
5. What are the basic molecular building blocks for fats?
6. Which foodstuffs are essential for the proper functioning of enzymes?
7. Name two compounds composed of many glucose molecules linked together.

Answers 1. Carbohydrates, fats and proteins. **2.** Fats. **3.** Enzymes, hormones, antibodies. **4.** Carbon, hydrogen, oxygen and nitrogen. **5.** Glycerol and fatty acids. **6.** Vitamins and minerals. **7.** Starch and cellulose.

Food tests

Key Fact
- A variety of chemical reagents can be used to test for different food groups.

Top Tip
The syllabus says you must know these four food tests. Remember the reagents and the colour changes.

The test for sugars

If the food you are testing is solid, then it must be crushed up with a little water to dissolve out the sugars.

Pour the liquid into a test-tube and add a few drops of blue-coloured Benedict's solution. Now heat the test-tube in a boiling water bath and if sugars are present the liquid will turn orange-brown. This is called the **Benedict's test**.

Unfortunately, the test does not work with all sugars. In fact, it works only with **reducing sugars**. These are sugars with particular chemical properties. Common table sugar, which is called **sucrose**, does not react with Benedict's solution.

Benedict's Solution

The test for starch

The test for starch does not involve any heat, and the food can be liquid or solid. A few drops of **iodine solution** are added to the food, and if starch is present the iodine changes from brown to a very dark blue-black colour. Try it with a piece of bread to see the test work well.

Iodine Solution

The test for protein

The test for protein does not involve any heat. If the food is solid it must be cut up into small pieces. Freshly made Biuret solution is added. If the Biuret solution turns from blue to lilac, then protein is present. This is called the **Biuret test**.

Try the test with milk to see the test work well.

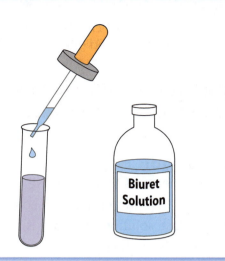

Biuret
Solution

The test for fat

Fats are not particularly easy to test for.
One test that works well with solids is to rub the food on a piece of brown parcel paper. If the food leaves a translucent mark then there is fat present. Try it with a piece of cheese to see the test work well.

Quick Test

1. What reagent is used to test for reducing sugars?

2. In which food test is heat required?

3. What is iodine solution used to test for?

4. If you tested a piece of potato with all four food tests, what results would you be likely to obtain?

5. Which line of the table shows correctly a food test reagent and the expected colour change?

	Reagent	Colour change
A	Benedict's	orange to blue
B	Iodine solution	orange to blue
C	Biuret	blue to lilac
D	Fat	blue to lilac

Top Tip
Never carry out food tests on your own and always follow your teacher's instructions.

Answers 1. Benedict's test. 2. Benedict's test. 3. Starch. 4. Positive for sugar and starch, negative for protein and fat. 5. C

Digestion

> **Key Fact**
> • Digestion involves the breakdown of large, insoluble food molecules into smaller soluble food molecules to allow absorption into the blood through the lining of the small intestine.

An experiment with digestion

Try the following experiment with the help of your teacher.

What you need:
- Two pieces of Visking tubing (The Visking tubing represents the lining of the gut.)
- Starch solution (The starch solution represents the food you eat.)
- Amylase solution (amylase is an enzyme that digests starch)
- One 250 ml beaker
- Two boiling tubes
- Iodine solution
- Benedict's solution

What to do:
- Wet one end of the Visking tubing and tie a knot in it.
- Add 5 cm³ of amylase solution to 20 cm³ of starch solution.
- Pour the mixture into the Visking tubing.
- Fold the other end of the Visking tubing over the lip of a boiling tube and hold it in place with an elastic band
- Fill the boiling tube with water at 37°C.
- Repeat the above but fill the tubing with 25 cm³ starch solution only (no amylase).
- Place this second tube in a separate boiling tube of fresh water.
- Leave the Visking tubing bags for as long as possible, at least 40 minutes, then test the water in each of the boiling tubes for starch and sugar.

Results
- Copy the table and write 'positive' or 'negative' in each of the four boxes.

	Starch test	Sugar test
Starch + amylase		
Starch on its own		

Conclusion
- See the Quick Test on the next page.

Quick Test

1. What is your conclusion?

2. Which tube acted as a control?

3. Why was it needed?

4. In this experiment, what does each of the following represent?
   - The Visking tubing
   - The starch
   - The boiling tube
   - The water in the boiling tube

Top Tip

Always try to do some experimental work to reinforce your knowledge.

Answers 1. The starch was unable to get through the Visking tubing wall. However, the amylase broke the starch down into sugar which was able to pass through the Visking tubing wall. **2.** The one without the amylase. **3.** To prove that it is the enzyme which breaks the starch down into sugar. **4.** Visking tubing = the intestine; starch = food that has been eaten; boiling tube = the body; water = the blood.

The need for digestion

Our gut is like a long tube running from one end of our body to the other. It is called the **alimentary canal**. If the food we eat is to be of any value to us it must get through the wall of the tube into our body. It can only do this if the food molecules are soluble and small enough to get through the tiny pores in the membranes of the cells lining our gut. Much of the food we eat is unable to do this; proteins, starches and fats have molecules that are insoluble and too large to get into the body. They have to be digested by enzymes which break them down to soluble products. Then the digested products are able to pass from the gut into the blood.

Top Tip

Some foodstuffs don't have to be digested. They can pass through the wall of the gut without any further treatment, e.g. salt, sugars and alcohol.

Quick Test

1. What are the end products of protein digestion?

2. What foodstuff does amylase digest?

3. Where does the food go after it has passed through the wall of the gut?

4. Name a foodstuff, which does not have to be digested.

Answers 1. Amino acids. **2.** Starch. **3.** Into the blood. **4.** Sugar, salt or alcohol.

Alimentary canal 1

Key Facts

- Food is broken down mechanically in the mouth and the stomach.
- Saliva contains amylase which digests starch to maltose.
- The stomach contains the enzyme pepsin which digests protein.
- The stomach contains acid which kills bacteria and softens bones.

The mouth

Food is broken up into small pieces by the action of the teeth. This makes the food easier to swallow and to digest. Every time a piece of food is cut in two a new surface is exposed to the action of enzymes. **Saliva** is added to the food to dissolve soluble substances and to make it easier to swallow. Moreover, saliva contains amylase which breaks down starch to maltose. Maltose is a sugar made up of two glucose molecules combined together. Later, the maltose will be broken down to glucose.

Top Tip
Try making flash cards with the biological name on one side and the description or definition on the other. Carry them with you to test yourself from time to time.

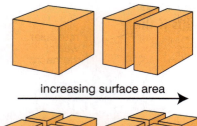

increasing surface area

Mouth contains teeth which begin the process by breaking up the food into smaller pieces

Salivary glands secrete amylase and mucus

Oesophagus, sometimes called the gullet

The oesophagus

The **oesophagus** is your throat; it joins the mouth to the stomach. It is a tube, but it is surrounded by muscle so that food and liquid is pushed down it rather than falling down it. In fact, it can work in reverse and does so when we are sick.

The muscles in the oesophagus are of two types; **longitudinal** and **circular**. The way they work can be seen in the diagram opposite.

This muscular process is called **peristalsis** and actually pushes food all the way along the gut, right to the end.

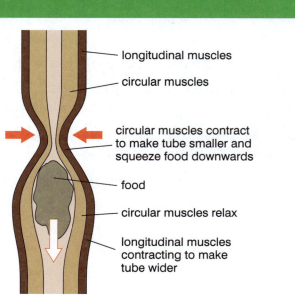

longitudinal muscles

circular muscles

circular muscles contract to make tube smaller and squeeze food downwards

food

circular muscles relax

longitudinal muscles contracting to make tube wider

The stomach

When the food reaches the stomach it is kept there for a few hours and mixed with **gastric juice** by the action of muscles in the stomach wall. Gastric juice is produced by special cells in the stomach wall. It contains hydrochloric acid, which is very effective at killing germs and at softening any bones that have been swallowed. Gastric juice also contains an enzyme called pepsin. Pepsin is unusual, in that it has become adapted to work in very acidic conditions (pH2). It digests protein to smaller molecules called **polypeptides**. These are later broken down to amino acids in the small intestine.

Gastric juice could easily digest the lining of the stomach as it is very acidic and breaks down protein. So the walls of the stomach are constantly protected by an alkaline slime (**mucus**) which is secreted by special cells in the stomach lining. Without this protective coating ulcers would develop very quickly. Mucus is produced throughout the digestive tract to help lubricate the passage of food and to protect the lining of the gut.

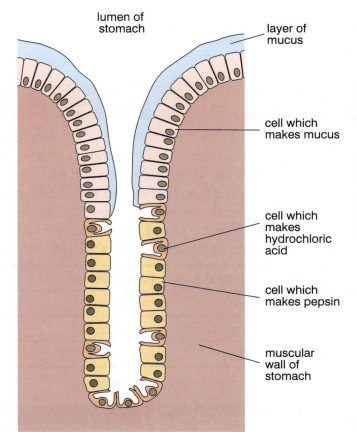

lumen of stomach

layer of mucus

cell which makes mucus

cell which makes hydrochloric acid

cell which makes pepsin

muscular wall of stomach

Quick Test

1. **a)** Name the enzyme found in saliva.

   b) What is its substrate?

2. What sugar is composed of two glucose molecules combined together?

3. What two parts of the body does the oesophagus link?

4. What name is given to the muscular process that pushes food along the gut?

5. What two kinds of muscle are found in the wall of the gut?

6. Name the two components of gastric juice.

7. Describe two functions of hydrochloric acid.

8. What does pepsin digest?

9. How does the stomach protect itself from self digestion?

Top Tip
Remember the three types of secretory cells found in the wall of the stomach.

Answers 1. a) Amylase. **b)** Starch. **2.** Maltose. **3.** Mouth and stomach. **4.** Peristalsis. **5.** Circular and longitudinal. **6.** Hydrochloric acid and pepsin. **7.** It kills bacteria and softens bones. **8.** Protein. **9.** By the production of mucus.

Alimentary canal 2

Key Facts
- The digestion and absorption of food takes place mostly in the small intestine.
- The intestine is lined with tiny projections called villi.
- Food is absorbed into the body through the walls of the villi.

The small intestine

When the food leaves the stomach it goes into the small intestine. There, **bile** and many enzymes are added to complete digestion. Some of the enzymes are produced by the walls of the intestine and others by the **pancreas**. The pancreatic juice and bile are both alkaline, to neutralise the acid of the stomach. The soluble products of digestion can then pass through the walls of the intestine into the **blood** and the **lymph system**. The table below summarises some of these digestive enzymes and their action.

Enzyme	Substrate	Products
amylase	starch	maltose
trypsin	polypeptides	amino acids
lipase	fats	fatty acids and glycerol

Absorbing food

The products of digestion diffuse through the membranes and cells of the intestine wall. Because this process is very slow, a large surface area is needed to ensure all the digested food can pass through into the body.

This is achieved in three ways:

- The small intestine is very long; around 4 to 5 metres.
- It is deeply folded.
- It is lined with tiny projections called **villi**.

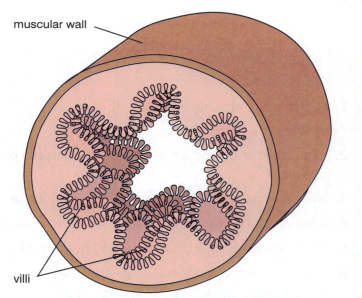

A cross-section of the small intestine

muscular wall

villi

Villi

Villi have a good blood supply and a central **lacteal**. This is a blind-ended tube that is part of the lymph system. The lymph system is a tubular system throughout the body that collects **tissue fluid** which has leaked out from the blood **capillaries**, and returns it to the blood near the heart.

All the products of digestion are absorbed into the blood capillaries except **fatty acids** and **glycerol**, which are absorbed into the lacteal. The blood flows from the capillaries to the **liver**; the lymph flows along lymph vessels to a point near the heart where it rejoins the main circulation.

Top Tip
Don't forget that the lacteal in a villus absorbs digested fats.

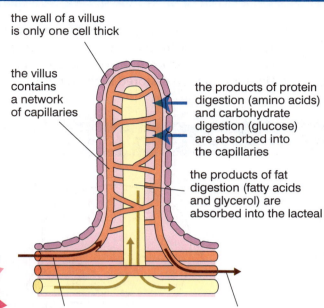

the wall of a villus is only one cell thick

the villus contains a network of capillaries

the products of protein digestion (amino acids) and carbohydrate digestion (glucose) are absorbed into the capillaries

the products of fat digestion (fatty acids and glycerol) are absorbed into the lacteal

blood arriving at the villus to pick up food molecules

blood leaving the villus, taking the food molecules to the rest of the body

The fate of food

- Glucose and other sugars are used in respiration to manufacture ATP. If too much sugar is eaten then it is stored as **glycogen** in the liver and muscles, or converted to fat and stored under the skin as we all know, sometimes to our cost.

- Fats have a similar fate, although they also have many other functions. Some act as hormones, they insulate us from the cold, and fats are a major component of all cell membranes.

- Proteins are used to make new body tissue, but they cannot be stored. Excess protein is broken down in the liver by a process called **deamination**. One of the products of this process can be used as a respiratory substrate, i.e. used in respiration to make ATP. The other product is **urea**.

Top Tip
Don't forget to learn the meaning of key words like deamination, villus, lacteal, lipase and trypsin.

Quick Test

1. What are the substrates of the following enzymes: lipase and trypsin?
2. What name is given to the blind-ended tube running down the centre of a villus?
3. Which of the following cannot be stored in the body: carbohydrates, fats or proteins?
4. What are the end products of protein digestion?
5. Where does deamination take place?
6. Where is glycogen stored in the body?

Answers 1. Fats and polypeptides. 2. Lacteal. 3. Proteins. 4. Amino acids. 5. The liver. 6. The liver and muscles.

Alimentary canal 3

Key Facts

- The pancreas produces enzymes for digestion.
- The liver deals with the products of digestion, which are carried there in the blood.
- Water is removed from the digested food in the large intestine.
- Undigested food is stored in the rectum.

Digestive system summary

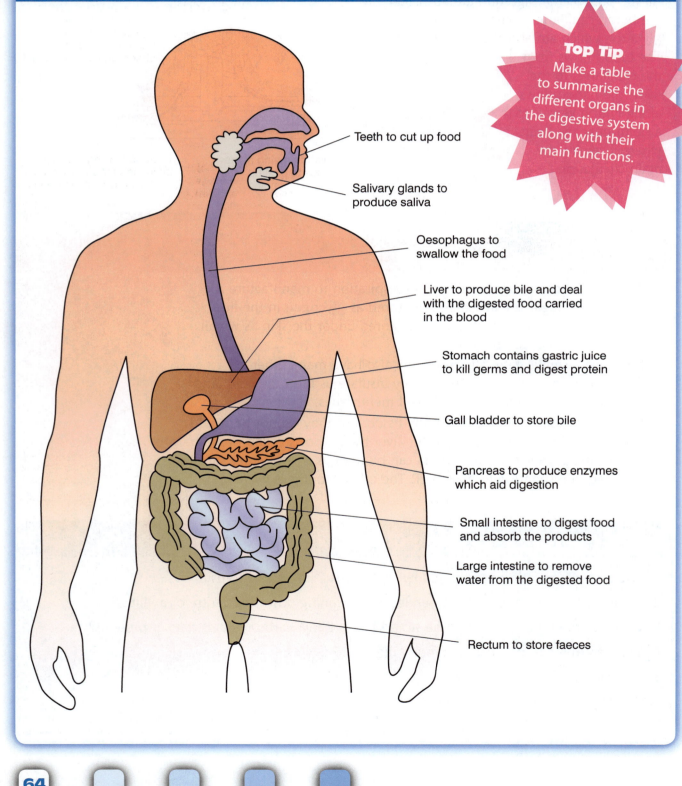

Top Tip
Make a table to summarise the different organs in the digestive system along with their main functions.

Teeth to cut up food

Salivary glands to produce saliva

Oesophagus to swallow the food

Liver to produce bile and deal with the digested food carried in the blood

Stomach contains gastric juice to kill germs and digest protein

Gall bladder to store bile

Pancreas to produce enzymes which aid digestion

Small intestine to digest food and absorb the products

Large intestine to remove water from the digested food

Rectum to store faeces

The pancreas

The pancreas is a small organ which lies just below the stomach. It is attached to the intestine by a duct that carries around a litre of pancreatic juice per day to add to the food in the intestine. The juice is alkaline, to neutralise the acid of the stomach, and it contains many enzymes including trypsin, lipase and amylase. The pancreas has another function too; it produces hormones which help regulate the sugar content of the blood.

The liver

The liver is like a chemical factory, carrying out thousands of different reactions every second. When the blood picks up the products of digestion in the intestine, it goes straight to the liver. There the liver evens out supply. If there is too much of any substance it will be stored or broken down. For example, excess glucose is stored as **glycogen** and excess proteins are **deaminated**. Glycogen is an insoluble carbohydrate rather similar to starch.

The liver has other functions too. Poisons, such as alcohol, are broken down, and red blood cells are destroyed. The iron in the red blood cells is recycled and other products from the red cell destruction are used in the manufacture of **bile**.

Bile

Bile is a greenish-yellow alkaline fluid produced in the liver and stored in the **gall bladder**. It contains a variety of substances, some of which give rise to the characteristic colour of our urine and faeces. It does not contain any enzymes, and its prime function is to break up fats into tiny droplets which are then easier to digest because of their increased surface area. This process is called **emulsification** and is rather similar to the action of washing-up liquid on the fats in a frying pan.

Top Tip
Don't forget that bile is not an enzyme and does not contain any enzymes.

The large intestine

The **large intestine** is shorter than the small intestine, but has a greater diameter, hence its name. When the undigested remains of the food reach the large intestine, much of the water is reabsorbed to leave a semi-solid product called **faeces**. The faeces are stored in the **rectum** and released from time to time through the **anus**.

Quick Test

1. Name two enzymes produced by the pancreas.

2. What happens to excess glucose in the liver?

3. Where is bile manufactured and stored?

4. Why is the emulsification of fats important?

5. What is removed from the digested food in the large intestine?

Answers 1. Trypsin, amylase or lipase. **2.** It is stored as glycogen. **3.** It is manufactured in the liver and stored in the gall bladder. **4.** It results in large fat globules being broken down into tiny droplets which are easier to digest. **5.** Water.

The kidney

Key Facts
- The kidneys treat the blood to remove poisons and to regulate its composition.
- They do this by two processes: filtration and reabsorbtion.

The urinary system

We have two kidneys. Each is supplied with blood via a **renal artery**. The urine which the kidneys produce is carried to the bladder in **ureters**. The bladder empties to the outside in both men and women via a small tube called the **urethra**.

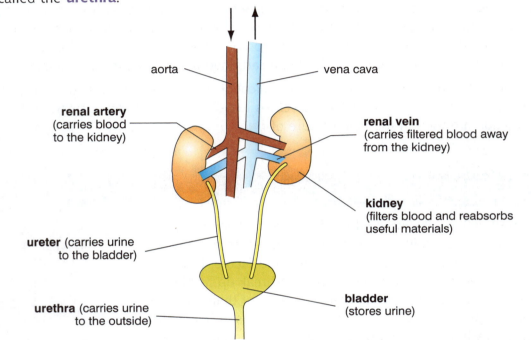

aorta

vena cava

renal artery (carries blood to the kidney)

renal vein (carries filtered blood away from the kidney)

kidney (filters blood and reabsorbs useful materials)

ureter (carries urine to the bladder)

urethra (carries urine to the outside)

bladder (stores urine)

Structure and function

The kidneys contain millions of tiny **filtering** units called **nephrons**. Each nephron has a cup-like structure called a **Bowman's capsule** which surrounds a tiny knot of blood capillaries called a **glomerulus**. Here the blood is filtered at a rate of around 180 litres every 24 hours.

The filters allow everything in the blood to pass into the **kidney tubules** except for blood cells and protein molecules, which are too large to pass through the microscopic pores. Ninety-nine per cent of the fluid that is filtered is reabsorbed by the blood. The reabsorbed filtrate contains useful substances like glucose and amino acids, leaving some water, excess salt, and poisons like urea to be carried to the bladder as urine. As a result of these two important processes your urine should not contain proteins, blood cells, glucose or amino acids.

Top Tip
Don't forget that urine is really only filtered blood and has nothing to do with the digestive system.

Urea

Excess proteins cannot be stored in the body. Instead, they are broken down in the liver by a process called **deamination**. Urea is one of the end products of this process and it is referred to as **nitrogenous waste** because it is mildly poisonous and contains the **nitrogen** that was present in the protein molecules. Urea is carried in the blood to the kidneys where it is removed.

Top Tip
Make sure you remember which substances are filtered and which are reabsorbed by the kidney tubules.

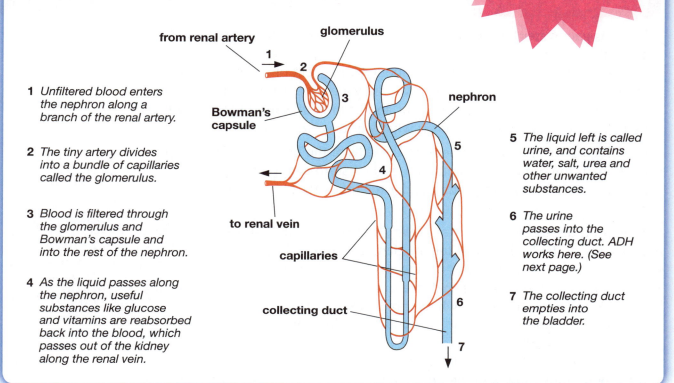

1 Unfiltered blood enters the nephron along a branch of the renal artery.

2 The tiny artery divides into a bundle of capillaries called the glomerulus.

3 Blood is filtered through the glomerulus and Bowman's capsule and into the rest of the nephron.

4 As the liquid passes along the nephron, useful substances like glucose and vitamins are reabsorbed back into the blood, which passes out of the kidney along the renal vein.

5 The liquid left is called urine, and contains water, salt, urea and other unwanted substances.

6 The urine passes into the collecting duct. ADH works here. (See next page.)

7 The collecting duct empties into the bladder.

Quick Test

1. Name the blood vessel that carries blood to the kidneys.
2. Name the tube that carries urine from the kidney to the bladder.
3. What two processes take place in a nephron?
4. Name a type of molecule that is too large to pass through the membranes of the Bowman's capsule.
5. Which of the following should not be present in the urine of a healthy adult: water, salt, protein, urea, glucose, red blood cells?
6. If 99 per cent of the filtrate is reabsorbed, using the data opposite, what volume of urine would be produced every 24 hours?

Answers 1. Renal artery. 2. Ureter. 3. Filtration and reabsorption. 4. Protein. 5. Protein, glucose and red blood cells. 6. 1% of 180 litres = 1.8 litres.

Osmoregulation

Key Facts
- The kidneys control the water/salt balance of the body.
- They do this under the influence of a pituitary hormone called ADH.
- Freshwater and marine fish have become adapted to their different watery environments.

Osmoregulation in mammals

Humans gain water through food and drink, and also as an end product of some chemical reactions that take place in the body. For example, water is an end product of respiration and about 10 per cent of our daily water requirement is met by this process. We lose water in our urine and faeces and in our breath and sweat. The salt content and volume of water in the body must be balanced at all times, and this is what we call **osmoregulation**.

Osmoregulation is important because the cells of the body would swell up or shrink if the osmotic concentration of their surroundings changed. The kidneys carry out osmoregulation under the influence of a hormone called **antidiuretic hormone (ADH)**.

ADH is produced by the **pituitary gland** which is situated on the underside of the brain. The pituitary gland is controlled by the **hypothalamus** which contains **osmoreceptors**. These are nerve cells that check the salt/water balance of the blood. ADH is carried to the kidneys by the blood, where it makes the **collecting ducts** of the nephrons more **permeable**. The collecting ducts are surrounded by salty tissue fluid, so water from the urine passes by osmosis back into the blood, making the blood more dilute and the urine more concentrated. When there is too much water in the blood, and the blood is more dilute, less ADH is produced and more water is lost in the urine because the walls of the collecting duct are no longer so permeable.

Top Tip
Learn when and how ADH works to regulate the water content of our bodies.

The Brain

hypothalamus

pituitary gland

Summary of what happens when you are thirsty:

Thirsty → hypothalamus stimulated → pituitary gland produces ADH → ADH travels to the kidney in the blood → kidney tubules become more permeable → more water is removed from the urine → the blood becomes more dilute and the urine becomes more concentrated.

Osmoregulation in freshwater fish

Freshwater fish such as pike, perch and sticklebacks tend to gain water by osmosis. Their scales are waterproof, but their gills are not, so water flows into their gills by osmosis because the concentration of salts in their body fluids is higher than that of the surrounding fresh water. As a consequence, these fish produce large volumes of dilute urine. In addition, because salts are in short supply, their gills absorb salts from the surrounding water.

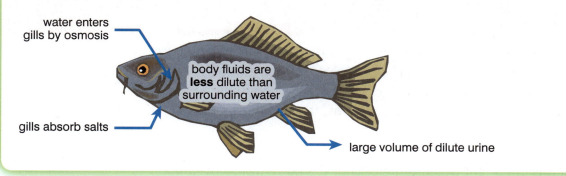

water enters gills by osmosis

body fluids are **less** dilute than surrounding water

gills absorb salts

large volume of dilute urine

Osmoregulation in marine (salt water) fish

Marine fish such as haddock, herring and cod tend to lose water by osmosis. Water flows out from their gills because the concentration of salts in their body fluids is lower than that of the surrounding salt water. Because of this, these fish drink salt water and produce only a small volume of concentrated urine. However, because they drink salt water, they have excess salts in their blood, which have to be secreted by special cells in their gills.

Top Tip
You should have spotted that the situation in freshwater fish and marine fish is exactly the opposite.

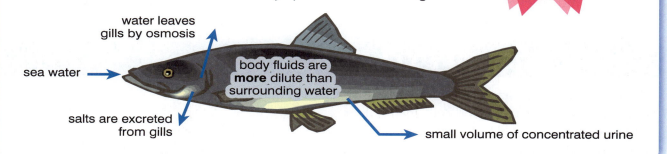

water leaves gills by osmosis

sea water

body fluids are **more** dilute than surrounding water

salts are excreted from gills

small volume of concentrated urine

Quick Test

1. Where are the osmoreceptors found in the body?

2. What effect does a shortage of water have on the pituitary gland?

3. What effect does ADH have on the collecting ducts?

4. What term is used to describe fresh water because it is more dilute than the body fluids of fish?

5. Compare the nature of the urine of freshwater fish with marine fish.

Answers 1. Hypothalamus. **2.** It is stimulated to produce ADH. **3.** It makes them more permeable. **4.** Hypotonic. **5.** Freshwater fish: high volume, dilute urine. Marine fish: low volume, concentrated urine.

Blood

> **Key Facts**
> - The blood is composed of a fluid, plasma, in which two types of cells are suspended.
> - Red blood cells carry oxygen and some carbon dioxide.
> - White blood cells protect the body from infection.

Plasma

The fluid portion of the blood is called **plasma**. It is a clear yellowish colour. The blood is the main transport system of the body, and the plasma contains many dissolved substances being carried to and from the body cells, e.g. hormones, sugars, salts, amino acids, proteins, fats, vitamins, urea and carbon dioxide.

The transport of carbon dioxide

Carbon dioxide, which is a waste product of respiration, is carried by the blood to the lungs where it diffuses into the air. It is carried in a variety of ways. Around 80 per cent of it is dissolved in the plasma, and the rest is carried by the haemoglobin in the red blood cells. The dissolved carbon dioxide makes the blood acidic and any change in pH is not tolerated by the body. So if the pH of the blood drops this stimulates us to breathe more quickly. Conversely, if we breathe too much, this can make the blood too alkaline and we feel dizzy.

Top Tip
Our breathing rate is influenced by the pH of the blood and not by the availability of oxygen.

The transport of oxygen

Oxygen is carried by **haemoglobin**. This is a special protein which gives red blood cells their colour. Red blood cells are not really true cells as they have no nucleus. Rather, they are just containers for haemoglobin. The characteristic doughnut shape of red blood cells is designed to maximise their surface area for gas exchange. When oxygen combines with haemoglobin it forms **oxyhaemoglobin**. Haemoglobin is very good at picking up oxygen in the lungs, and it is also very good at releasing oxygen in the tissues where there is a shortage of oxygen.

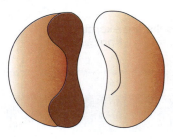

this diagram shows a red blood cell that has been sectioned to show its characteristic shape

White blood cells

There is only one white blood cell for every 600 red blood cells in the body. However, they have a very important function; they protect the body from invading microbes.

There are two major classes of white blood cells: **macrophages** and **lymphocytes**.

Macrophages wander throughout our body tissues engulfing invading or infected cells by a process called **phagocytosis**. Once ingested, the cells are digested by powerful enzymes.

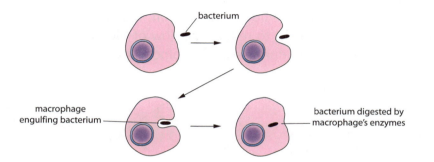

bacterium

macrophage engulfing bacterium

bacterium digested by macrophage's enzymes

Lymphocytes produce **antibodies**. These are special proteins that attach themselves to invading viruses and bacteria and ensure they are destroyed. Antibodies are like enzymes in that they are highly specific; only one antibody can attack one kind of microbe. So we have thousands of different lymphocytes in the blood and throughout the tissues, each producing specific antibodies when required.

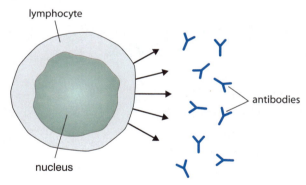

lymphocyte

antibodies

nucleus

Top Tip

Make sure you remember the ways macrophages and lymphocytes protect the body.

Quick Test

1. What name is given to the fluid portion of the blood?
2. Name three substances dissolved in the plasma.
3. How is carbon dioxide carried in the blood?
4. What effect does carbon dioxide have on the pH of the blood?
5. What kinds of cells produce antibodies?
6. What kinds of cells carry out phagocytosis?

Answers 1. Plasma. 2. Carbon dioxide, glucose, amino acids. 3. In the plasma and in the red blood cells. 4. It lowers it. 5. Lymphocytes. 6. Macrophages.

Blood vessels

> **Key Facts**
> - Blood is carried around the body in arteries, veins and capillaries.
> - Arteries carry blood away from the heart and veins carry blood towards the heart.
> - Capillaries join arteries to veins and are permeable to allow exchange of materials with body cells.

Arteries

Arteries have thick elastic walls because they have to withstand the high pressure of blood which has been pumped by the **heart**. In addition, they have circular muscles in their walls, which can constrict to reduce the flow of blood or relax to let it flow more easily. For example, during exercise, blood flow is diverted from the gut to the muscles, and when we are warm, blood flow is increased to the surface of the skin.

You can feel arteries through your skin because they pulsate with the beating of the heart. They are deep within the body to protect them from damage, because if cut, they would lose blood very rapidly.

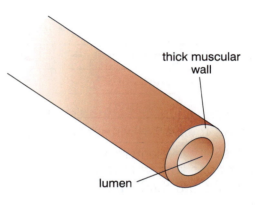

thick muscular wall

lumen

Veins

Veins carry blood from the tissues back to the heart. They are thin-walled with very little muscle. The veins carry blood at low pressure because it has lost most of its pressure flowing through the millions of narrow capillaries. In fact the pressure is so low that the blood would not return to the heart without the assistance of **valves**. When we move our body we squeeze and bend our veins. This pushes the blood along them, but the blood can only flow one way because the valves stop it flowing backwards.

Top Tip
Use mnemonics to remember facts, e.g. **A**rteries = **A**way from heart and **V**eins = **V**alves

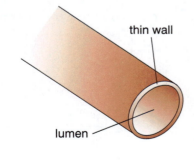

thin wall

lumen

longitudinal section of vein

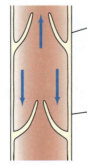

valves open as blood flows through

valves close to prevent back-flow of blood

Capillaries

Capillaries are tiny and have walls only one cell thick. Unlike arteries and veins, they are permeable. Plasma leaks through the walls of the capillaries, carrying useful substances with it. This leaked tissue fluid returns to the blood eventually, much of it being picked up by the lymph vessels and returned to the veins near the heart.

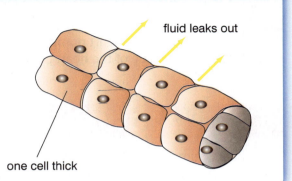

fluid leaks out

one cell thick

The main blood vessels of the body

| Organ | Blood supply | |
	going to	away from
Liver	Hepatic artery	Hepatic vein
Kidney	Renal artery	Renal vein
Lungs	Pulmonary artery	Pulmonary vein
Intestines	Mesenteric artery	Hepatic portal vein
Heart	Coronary artery	Cardiac vein

Top Tip
Note that the liver has a dual blood supply. The hepatic artery brings oxygenated blood and the hepatic portal vein brings blood with digested food from the gut.

Quick Test

Use the words artery, vein or capillary to answer the following questions:

1. Which one pulsates?

2. Which one has valves?

3. Which one has thick muscular walls?

4. Which one is permeable?

5. Which one carries blood to the heart?

6. Which one carries blood at the lowest pressure?

7. Which one can change its shape to control the flow of blood?

Top Tip
Learn the names of the blood vessels in the table because they are all mentioned in the examination syllabus.

Answers 1. Artery. **2.** Vein. **3.** Artery. **4.** Capillary. **5.** Vein. **6.** Vein. **7.** Artery.

The heart

> **Key Facts**
> - The heart is a double muscular pump.
> - One side of the heart pumps blood to the lungs.
> - The other side of the heart pumps blood to the rest of the body.

The function of the heart

The heart pumps blood around the body. We need it because diffusion is too slow a process on its own for carrying substances to and from our cells. The heart normally beats at around 70 beats per minute, but can be as slow as 40 beats per minute in a resting athlete, or beat at over 200 beats per minute in a young person undergoing strenuous exercise.

Each time the heart beats, valves close to stop the blood being pushed backwards. There are two pairs of valves which close one after the other, giving the lub-dub sound of the heart beat.

The structure of the heart

The heart has four chambers and four valves. The top two **atria** are collecting chambers that pump the blood into the chambers below. The **ventricles** pump blood around the body, so they are much stronger.

The heart is really composed of two pumps combined together as one. When the blood comes back from the body to the heart it needs to get rid of excess carbon dioxide and pick up oxygen, so the heart pumps it to the lungs. However, by the time the blood has been through the lungs it has lost pressure and there is not nearly enough pressure left in the system to push the blood around the rest of the body. So the blood returns to the heart for a boost before setting off on its journey round the body.

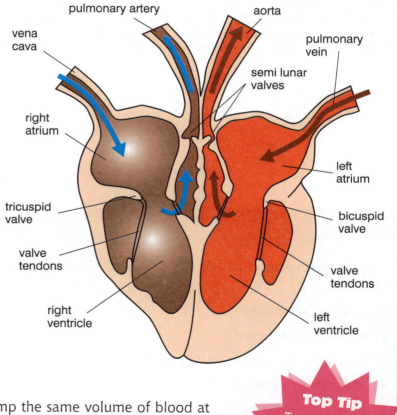

The two pumps work together and pump the same volume of blood at each beat. The pressure in the **pulmonary** system (the one going to the lungs) is much lower because not a great deal of pressure is needed to pump blood round the lungs. Consequently, the muscular walls of the right ventricle are thinner and less strong than those on the left.

Top Tip
Try to work out which valves close and open as the heart beats.

The valves of the heart

There are two pairs of valves in the heart and their job is to stop the backflow of blood in the heart.

When the atria conract to fill the ventricles, the triscupid and bicuspid valves open to allow the blood to flow through to the ventricles. When the ventricles contract to pump blood to the body, the semilunar valves open to allow the blood to flow out of the heart and the tricuspid and bicuspid valves close. When the ventricles relax, the semilunar valves close to stop the blood flowing back into the ventricles.

The blood supply to the heart

Although the heart pumps blood, the heart muscle itself needs its own blood supply. It gets this from the **coronary artery** which is attached directly to the aorta. If you do not take enough exercise, live a stressful life and have a diet rich in fat, then you risk damaging these arteries. The lining of the coronary arteries becomes clogged with a fatty substance and the blood supply to the heart muscle becomes restricted. At worst, arteries can become totally blocked and a person then suffers from a heart attack (a coronary thrombosis) and may even die.

Top Tip
Eat a healthy diet and take plenty of exercise to give your heart a chance.

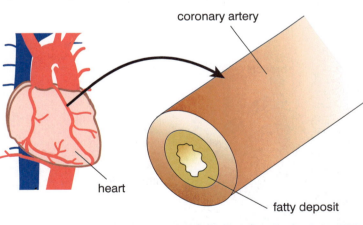

coronary artery

heart

fatty deposit

Quick Test

1. Name the heart chamber that receives blood from the body.
2. Which side of the heart deals with deoxygenated blood?
3. In which chamber of the heart is blood pressure greatest?
4. Name the valve that separates the right atrium from the right ventricle.
5. What happens to the coronary arteries if you don't look after your heart?

Answers 1. Right atrium. 2. Right. 3. Left ventricle. 4. Tricuspid valve. 5. They become blocked.

The lungs

Key Facts
- The lungs are designed to be efficient at gas exchange.
- They are composed of many tubes, each ending in microscopic air sacs called alveoli.

The function of the lungs

The lungs ensure that we obtain a supply of oxygen for aerobic respiration, and they allow us to get rid of the waste product carbon dioxide. Again, the process of diffusion is too slow to supply our needs, so we have muscles in our chest which suck in and blow out the air much more quickly than it would diffuse.

The structure of the lungs

The lungs are not like two balloons in our chest. They are composed of millions of tiny air sacs called **alveoli**. The subdivision of one air space into millions of tiny air spaces results in a huge increase in surface area. Connected to the air sacs are millions of tiny air passages called **bronchioles**. These join to form **bronchi** (singular bronchus) which finally link to make a single tube in our neck called the **trachea**. The trachea and bronchi are held open by bands of **cartilage**, a tough, springy material. Bronchioles have no cartilage and are held open by protein fibres and rings of muscle, which can contract or relax and hence affect air flow through the bronchioles. When these muscles contract in spasm, a person has an asthma attack.

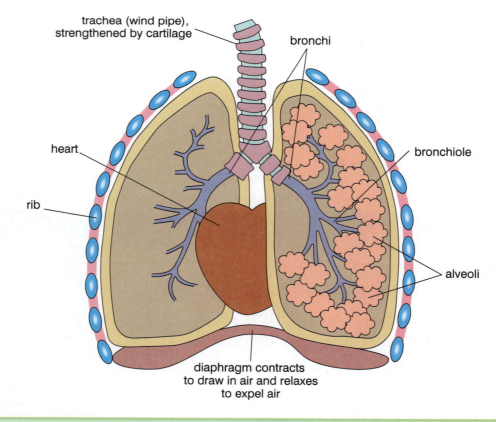

trachea (wind pipe), strengthened by cartilage

bronchi

heart

bronchiole

rib

alveoli

diaphragm contracts to draw in air and relaxes to expel air

Alveoli

The alveoli are well adapted for gas exchange:

- They have very thin membranes so the gases do not have far to diffuse.
- They have a rich blood supply.
- They have a moist lining, which allows gases to dissolve in the body fluids.
- Taken together, they have a very large surface area.

Top Tip
Remember the list of features of the alveoli that ensure efficient gas exchange.

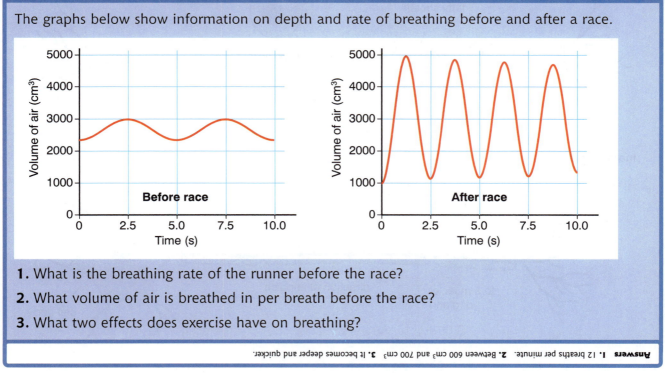

Quick Test

The graphs below show information on depth and rate of breathing before and after a race.

1. What is the breathing rate of the runner before the race?

2. What volume of air is breathed in per breath before the race?

3. What two effects does exercise have on breathing?

The nervous system

Key Facts

- The nervous system consists of the brain, the spinal cord, and the nerves that link these two organs to the rest of the body.
- The nerves carry messages from the brain to various organs in the body and vice versa.
- Reflex actions are quick, automatic and difficult to suppress.

The central nervous system (CNS)

The **brain** and **spinal cord** make up the **CNS**. They contain millions of **motor neurones** (nerve cells) that send messages to the muscles and the glands of the body, and **sensory neurones** that receive information from our surroundings. Our skin, eyes, nose, ears and tongue are the sense organs that pick up information and send it to the brain. If the brain decides to respond, nervous impulses will be sent to glands or to muscles to bring about an appropriate response. Because of this, we call muscles and glands **effectors**.

Top Tip
Don't forget that the CNS is composed only of the brain and spinal cord.

Nerve cells

Nerve cells (**neurones**) are cells that have many fibres, some as long as a metre. These connect with one another and with muscles, glands and the sense organs of the body. Nervous impulses are passed along the fibres to and from the brain.

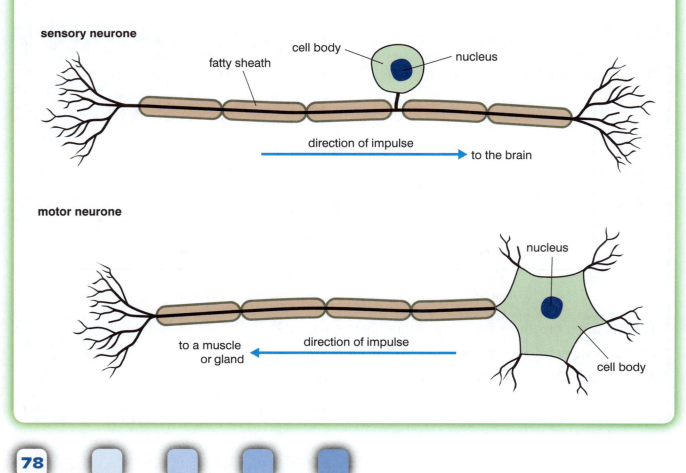

sensory neurone

cell body · nucleus · fatty sheath · direction of impulse · to the brain

motor neurone

nucleus · to a muscle or gland · direction of impulse · cell body

Reflex actions

Reflex actions are automatic, rapid, simple responses to stimuli. They are also difficult to suppress. They include coughing, sneezing, blinking and withdrawing your hand from a very hot object. We are born with these responses, and do not have to learn them, which is just as well, because most of them are protective in nature.

The diagram below shows how a **reflex arc** works. The key feature of reflexes is that the body will react directly, before the brain has time to make a conscious decision. This is because the incoming nervous impulse is short-circuited straight across the spinal cord to the outgoing motor neurone, which stimulates the muscle to contract a fraction of a second before the brain is aware of the problem.

Top Tip
Remember the important features of reflex actions.

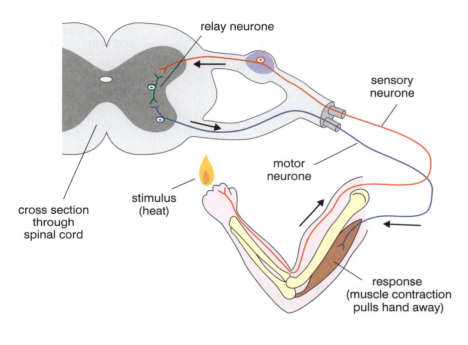

relay neurone

sensory neurone

motor neurone

cross section through spinal cord

stimulus (heat)

response (muscle contraction pulls hand away)

Quick Test

1. What do the letters CNS stand for?

2. Name the two principal types of neurone.

3. Which of these two links the brain with the muscles?

4. Describe three features of reflex actions.

5. Give two examples of reflex actions.

6. What is an effector?

Answers 1. Central nervous system. **2.** Sensory and motor. **3.** Motor. **4.** They are quick, automatic and difficult to suppress. **5.** Coughing and sneezing. **6.** A muscle or gland which responds to impulses from a motor neurone.

The brain

Key Facts
- The brain is composed of three principle regions: the cerebrum, the cerebellum and the medulla oblongata.
- The brain has discrete areas controlling different functions, such as the motor and sensory strips.

The brain

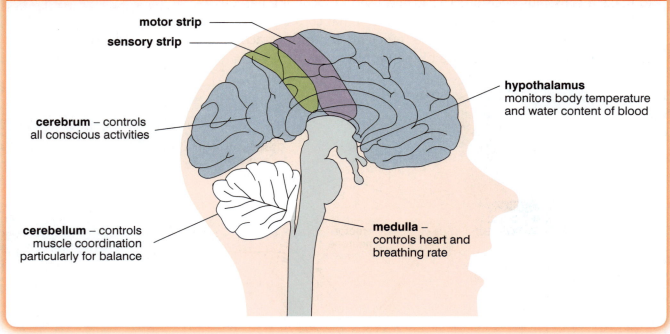

motor strip
sensory strip

hypothalamus
monitors body temperature
and water content of blood

cerebrum – controls
all conscious activities

cerebellum – controls
muscle coordination
particularly for balance

medulla –
controls heart and
breathing rate

The cerebrum

The cerebrum is the largest part of the brain and it is the one we are most aware of because it dictates our conscious activities. It is composed of two halves, called the cerebral hemispheres. The front of the cerebrum is the centre of our emotions and personality. Further back, the sensory strip and motor strips receive and send information from and to all parts of the body. At the back of the brain, information from our eyes enables us to see. Many parts of the cerebrum are involved in memory, and all our speech is co-ordinated and controlled by the cerebrum.

The sensory and motor strips

The sensory strip deals with all incoming information from the senses: the ears, the eyes, the nose, the tongue and the skin. The motor strip sends information to all the muscles of the body that we control by conscious thought. Not surprisingly, the area of the strip assigned to each part of the body is proportional to the amount of information received or sent, e.g. a large proportion is devoted to the hands.

Top Tip
The motor strip is the one towards the front of the brain, just as the motor is at the front of a car.

The cerebellum

The **cerebellum** controls subconscious activities associated with the fine motor control of muscles, especially in maintaining our balance. So when we write, walk, swim or catch a ball we are using our cerebellum to co-ordinate our muscles in a sophisticated way.

The medulla oblongata

The **medulla** controls the most basic bodily functions needed to maintain life, e.g. heart rate, breathing rate, digestion and some reflex actions. When the medulla ceases to function doctors will declare us dead.

Top Tip
As you work up from the base of the brain to the front of the brain you go from basic functions to the most complex of functions which make us uniquely human.

The hypothalamus

The hypothalamus is situated under the brain, in close contact with the pituitary gland. It is connected by nerve fibres to most other parts of the nervous system, and despite its small size, its functions are bewilderingly diverse. It affects our sleeping patterns, gives rise to feelings of pleasure, fear and rage, and to the sensations of thirst and hunger. It also monitors temperature and the osmotic concentration of the blood.

Top Tip
The syllabus says you should know the hypothalamus is the centre for the regulation of water balance and temperature.

Quick Test

Use the words cerebrum, cerebellum, medulla and hypothalamus to complete the following:

1. Centre for conscious thought.
2. Needed for co-ordinating our muscles for walking.
3. Controls sneezing and coughing.
4. Monitors our body temperature.
5. Contains the motor and sensory strip.
6. Regulates our heart rate.

Answers 1. Cerebrum. 2. Cerebellum. 3. Medulla. 4. Hypothalamus. 5. Cerebrum. 6. Medulla.

Temperature regulation

Key Facts

- The body temperature of mammals and birds is kept within a very narrow range.
- The hypothalamus monitors body temperature.
- There are many conscious and subconscious mechanisms that help us maintain a constant body temperature.

Why temperature control?

Mammals and birds are particularly good at maintaining a stable body temperature. This keeps enzyme activity very constant and enables these animals to exploit regions of the world that cold blooded animals find inhospitable. The down side is that birds and mammals have to eat very large quantities of energy-rich food to fuel their central heating systems. The ability to maintain a constant temperature involves a number of **negative feedback** mechanisms. (See page 83.)

Top Tip
Remember that a constant body temperature ensures that there is an ideal environment for the activity of enzymes.

Keeping warm

When the body temperature drops, mammals will start to shiver. This is the spasmodic contraction of muscles to generate heat. Sweat glands close and the hairs on the skin are raised to trap air because air is a very good insulator. Also, blood flow to capillaries in the skin decreases as it is diverted to deeper blood vessels. This is called **vasoconstriction** and makes us look pale in appearance. In the long term, our metabolic rate increases and we may find that we have to eat more food to keep warm. In addition, many mammals have layers of thick hair or fat to insulate them from cold conditions.

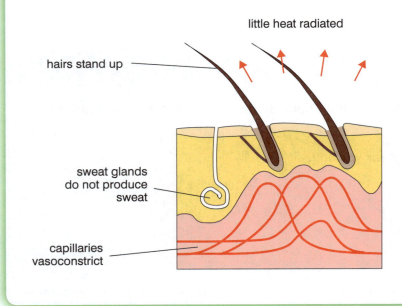

little heat radiated

hairs stand up

sweat glands do not produce sweat

capillaries vasoconstrict

Keeping cool

If our body temperature rises, many of the mechanisms described opposite are reversed. The hair in our skin lies flat, the capillaries **vasodilate** and we appear flushed. In addition, sweat glands produce sweat. Sweat in itself is not cold. It is the evaporation of sweat from the skin that draws heat from the body. In very humid conditions where sweat cannot evaporate, this cooling mechanism is of little value. Our metabolic rate will decrease over the long term and our demand for food will decline.

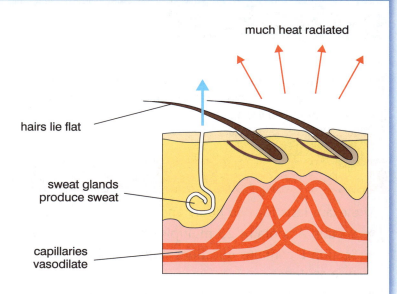

much heat radiated

hairs lie flat

sweat glands produce sweat

capillaries vasodilate

Conscious activities

All the responses described above are automatic. However, humans can make many conscious decisions to help maintain a comfortable body temperature. We can seek shade or shelter, curl up or stretch out, change the clothes we wear and switch on the heating or air conditioning.

Negative feedback control

You have learned in the last few pages about systems that maintain a constant temperature and water content of the body. They all rely on a control mechanism called **negative feedback**. When a factor increases beyond a set point, it triggers a response that causes it to decrease and return to the set point, and vice versa. Many simple control systems in your house work in the same way, e.g. thermostats. When you set a thermostat it switches on the heating when it gets too cold and switches it off when it gets too hot.

Top Tip
Make sure you can describe how negative feedback mechanisms work in relation to the control of temperature and water balance.

Quick Test

1. What happens to capillaries near the surface of the skin when we are too hot?

2. What must happen to sweat before it has a cooling effect?

3. Which chemicals in our cells are particularly sensitive to temperature change?

4. What term is used to describe the control mechanism that maintains body temperature?

Answers 1. More blood is diverted through them to enable the body to lose more heat. 2. It must evaporate. 3. Enzymes. 4. Negative feedback control.

The exam and the syllabus

Key Facts
- The examination is based on the syllabus, which can be obtained from SQA.
- You must know the facts, but you must also be able to solve problems and comment on practical work.

The exam

The exam lasts for two hours and offers a maximum of 100 marks

- Section A contains 25 multiple-choice questions worth 1 mark each.
- Section B contains questions requiring short answers of no more than a word or a sentence.
- Section C contains two short essay-type questions worth 5 marks each.

You will be awarded an A, B, C or D depending on your mark. The SQA tries to keep the cut-off scores as close to 70, 60, 50 and 45 for A, B, C and D respectively.

You can find the date of the exam by going to the SQA website. The dates of the exams are published one year in advance under the heading 'Examination Timetable'.

The examination questions are set from information given in the syllabus and not from textbooks.

The unit tests

You can sit the examination on its own and you will be credited with a mark and a grade. However, to gain the full award you must sit three unit tests and write up a practical investigation, which you have carried out on your own in the laboratory or as fieldwork.

Unit tests are sometimes called NABs. NAB stands for National Assessment Bank. The tests are worth 40 marks and you must gain at least 26 marks to pass. If you fail, you may try again with a different test. Your teacher will give you this test and mark it, but the mark is only used to award a pass or fail. Note that the unit headings in this book relate to the three tests that you must sit.

The overall grade you obtain in the final exam is not affected by your performance in the NAB tests or your practical work.

The syllabus

This is a document published by the SQA which tells teachers, examiners, candidates and textbook writers what should be taught and examined.

The syllabus states very clearly what you need to know. In particular it has three columns, and the examination is based on the first two columns called CONTENT and NOTES.

You can obtain your own copy in the following way:

- Go to www.sqa.org.uk
- Click on Pupil
- Click on Biology
- Click on Intermediate 2
- Click on Arrangements Documents
- Print out pages 7 to 29. Don't print out the whole publication because it is 64 pages long!

Top Tip
It is definitely worth getting a copy of the syllabus from your teacher or from the Internet.

This book

This book is based exclusively on the syllabus, so you can match each page to the content statements in the syllabus. It gives you a description of all the biological terms you need to know in the glossary. The glossary is based on the syllabus and on terms appearing in past exam papers, so it is worth committing to memory. Use flash cards to help you with this. Cut out a piece of card, copy the word on one side and the definition on the other. Carry the cards about with you and test yourself from time to time.

In addition, try all the questions at the foot of each page. Repeat these a few times until you get them all correct.

Past papers

One of the best ways to become familiar with the exam is to do past papers. These are produced by Leckie & Leckie every year and are available in many bookshops. They include the latest five SQA examinations with full sets of answers. It's well worth buying a copy and working your way through the questions. By using Leckie & Leckie's Official SQA Past Papers, along with this textbook, you should have no difficulty passing the Unit tests and the final examination.

Top Tip
When doing past papers, never be tempted to look up the answer until after you have made a really good attempt at answering the question.

Passing the exam

Key Facts
- Read the questions carefully and respond appropriately.
- Practice makes perfect.

Short-answer questions

- Try to spell words correctly. The markers will allow some errors, but not many.
- The following abbreviations are acceptable in your answers: DNA, ATP and ADH.
- Keep your hand-writing clear. You will never be able to meet the marker to tell him or her what you meant!
- If asked for a description use a sentence or two to state what it is or what it does.
- If asked for an explanation use a sentence or two to say why or how something has happened.
- Never answer requests for descriptions or explanations with one word.
- If asked to carry out a calculation, remember to give the units if they are not already given, and to use the decimal place correctly, e.g. $4\cdot567$ cm^3 to one decimal place is $4\cdot6$ cm^3.
- If the word 'rate' is used, remember that time is involved, e.g. $4\cdot6$ cm^3 of gas per minute.
- Check the number of marks for a question. If there are 2 marks then you need to provide at least two pieces of information.
- Go through the paper answering the questions you find easy and skipping the questions you find difficult. Go back to the most difficult questions towards the end of the exam. This ensures you gain maximum marks quickly and it raises your confidence.

Top Tip
Remember the difference between 'describe' and 'explain'.

Quick Test

An investigation was set up to check the effect of caffeine on heart rate in humans. Twenty subjects had their heart rates measured before and after drinking a cup of caffeinated coffee.

1. **Describe** a control for this investigation.

2. **Explain** the need for this control.

Answers 1. Twenty other subjects should be treated in exactly the same way but given decaffeinated coffee instead. (Ideally no-one should know what kind of coffee they are drinking.) **2.** The control is there to show that it is the caffeine in the coffee which is having the effect and nothing else present in the coffee.

Multiple choice questions

There are 25 multiple choice questions (MCQs) at the beginning of the exam paper, each worth 1 mark. They cover any topic and some of them test your problem solving and practical abilities.

The multiple choice questions are at the beginning of the exam paper for a reason. They are generally the easiest part of the exam, because the answers are given to you and because you do not have to write anything, just mark a box. Consequently, always try this paper first as you will gain marks quickly and you are likely to gain confidence too. However, don't forget, some of the MCQs are difficult and it may take time to reach an answer.

Never leave an MCQ unanswered. Always guess, even if you don't have any idea, because your guess might be correct.

Good examiners always ensure there is a more or less even spread of answers, and that there are never any long sequences of the same answers, e.g. B B B B B B C A D B is unlikely to turn up. So, be suspicious if you find you have too many of one letter in your answers.

Work out the answer yourself before looking at the four possible answers.

Top Tip
You have to use a pencil and rubber to answer the MCQ paper so that mistakes are easy to correct.

Examples of MCQs

1. Which of the following are the end products of anaerobic respiration in yeast?

    A carbon dioxide and water

    B carbon dioxide and ethanol

    C ethanol and pyruvic acid

    D pyruvic acid and lactic acid

2. Which part of the brain monitors the temperature of the blood?

    A Cerebellum

    B Hypothalamus

    C Pituitary gland

    D Medulla oblongata

3. Which factor is limiting the rate of photosynthesis at point X on the graph?

    A Carbon dioxide concentration

    B Temperature

    C Light intensity

    D Oxygen concentration

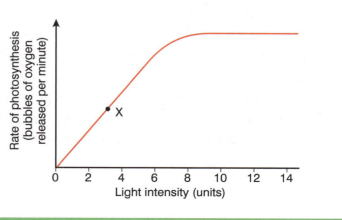

Answers 1. B 2. B 3. C

Problem solving and practical abilities

Key Facts
- The exam tests knowledge (KU), problem solving (PS) and practical abilities (PA).
- There will be between 9 and 11 questions on PS and PA in the MCQ paper.
- There will be between 15 and 20 marks covering PS and PA in the main paper.

What the syllabus says about PS and PA

Candidates should be able to:
- select information from text, tables, graphs, charts, keys and diagrams
- present information in a variety of forms, including written summaries, tables and graphs
- calculate percentages, averages and ratios and use significant figures appropriately
- plan and design procedures to test hypotheses
- identify and understand the need for controls
- identify variables that should be controlled in experimental situations
- evaluate unfamiliar experiments by commenting on the suitability of procedures, controls, control of variables, limitations of equipment and possible sources of error
- draw valid conclusions from unfamiliar experimental situations
- describe overall patterns or trends in readings or observations
- make predictions based on evidence.

Top Tip
Examiners are instructed to cover all ten PS and PA points in every exam.

Graphs

- There will always be graphs of one kind or another in the exam.
- When taking readings from a graph remember to state the units.
- When drawing a graph:
  - complete the *x* and *y* axes with numbers and units
  - make sure the numbers rise in equal increments, e.g. 5, 10, 15, 20
  - join the points with straight lines using a ruler
  - choose a scale that uses up most of the graph paper
  - never plot to zero unless you have been given data for zero
  - make sure you have drawn the right kind of graph: a line or bar graph.

Percentages

A simple rule that works for percentage calculations is to find out the change that has taken place and divide this by the starting value, then multiply by 100.

i.e. $\dfrac{\text{Change}}{\text{Starting value}} \times \dfrac{100}{1}$

A population of mice increased from 250 to 300 in one month. What is the percentage increase?

Change = 50. Starting figure = 250. So percentage increase is $\dfrac{50}{250} \times \dfrac{100}{1} = 20\%$

Ratios

You will be given two or three numbers and asked to express them as a simple whole-number ratio.

For example:

The following species of tree were found in an area of woodland: 55 Ash, 10 Oak and 20 Beech trees.

- Look for a number that can divide each of these numbers and still leave a whole number.
- If you divide by 10 you get 5·5, 1 and 2, but 5·5 is not a simple whole number, so this is wrong.
- If you divide by 5 you get 11 and 2 and 4, so this is the correct answer.

Top Tip
Make sure you can calculate percentages, averages and ratios.

Hypotheses

A hypothesis is a predicted outcome of an investigation based on previous experience or knowledge.

Reliability

The best way to improve reliability is simply to repeat the investigation to obtain more readings. This will give you more reliable averages which reduce the impact of unusual results. (See pages 19 and 35.)

Validity

If you are investigating the activity of an enzyme at different temperatures you must keep all other factors the same. If you don't, your investigation is invalid because you don't know which factor has caused a change. Investigations are either valid or invalid, there is no intermediate state. (See page 35.)

Control of variables

You should keep all variables (factors that can be altered) constant, except the one you are testing. If you allow two variables to change then your experiment becomes invalid. (See page 35.)

Controls

Check pages 19, 59 and 86 to see what is meant by a control and why it is important in many experiments.

Top Tip Be sure you know the difference between reliable and valid, and between controls and control of variables.

Patterns and trends

When describing patterns or trends shown by a graph it is a good idea to quote data from the graph, e.g. the graph on page 16 could be described as follows: the reaction rate rises up to a maximum at 40°C and then declines.

MCQs which test Practical Activities

1. The diagram shows apparatus used in an investigation into the production of apple juice using cellulase enzyme.
   Which of the following would be a suitable control?
   The same apparatus with

   A no apple puree

   B no cellulase

   C apple puree replaced with water

   D cellulase replaced with water

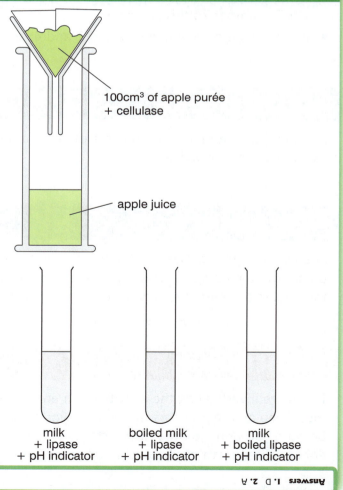

100cm³ of apple purée + cellulase

apple juice

2. Three test-tubes were set up as shown, to investigate the activity of lipase.
   How could the experiment be made more reliable?

   A By repeating it a few times and averaging the results

   B By using boiled milk in all the test tubes

   C By keeping the temperature the same

   D By using different types of milk

milk + lipase + pH indicator

boiled milk + lipase + pH indicator

milk + boiled lipase + pH indicator

Answers 1. D 2. A

Glossary

Top Tip
The glossary is a list of all the words you should know for passing the exam.

activation energy   the energy input required to start chemical reactions

active site   part of an enzyme molecule where a catalytic reaction takes place

ADH   antidiuretic hormone; produced by the pituitary in response to thirst

ADP   adenine diphosphate; combines with phosphate to make ATP

aerobic respiration   a process using oxygen, which releases energy from food, to make ATP

alimentary canal   the digestive tract from mouth to anus

allele   an alternative of a gene; e.g. tall and dwarf are alleles for height in peas

alveolus   a tiny thin-walled air sac in the lung; lungs contain millions of them (alveoli)

amino acid   the basic building block for the manufacture of proteins

amylase   an enzyme that digests starch to form the sugar maltose

anaerobic respiration   a process that releases energy from food without the need for oxygen

anther   part of a flower where pollen, containing the male gamete, is produced

antibiotic   a compound that kills bacteria; first found in fungi

antibody   a protein produced by lymphocytes to defend against bacteria or viruses

anus   the opening at the end of the alimentary canal

aorta   the main artery carrying blood from the heart to the body

artery   a blood vessel that carries blood from the heart to body organs

ATP   a high-energy compound that releases energy to drive metabolic reactions

atrium   an upper chamber of the heart that receives blood from a vein

bacterium   a one-celled organism, much smaller than an average cell, with no nucleus

base (DNA)   a molecular sub-unit of a nucleotide, of which there are four different ones

Benedict's test   a test for sugars in food in which the heated liquid turns from blue to brown

bile   an alkaline fluid produced by the liver, which emulsifies fats

biodiversity   the number and variety of different species in any ecosystem

biogas   a mixture of methane and CO_2 produced by decomposing bacteria

Biuret test   a test for protein in food in which the liquid turns from blue to lilac

Bowman's capsule   part of a kidney nephron that collects filtrate from the glomerulus

brewing   the fermentation of sugar by yeast to produce alcoholic drinks

bronchiole   a small tube which carries air in the lungs

bronchus   a large tube, with rings of cartilage, which carries air in the lungs

capillary   a tiny blood vessel, which is permeable to allow for exchange of materials

carbohydrate   a compound containing C, H and O, and used as a source of energy

carbon dioxide   a gas produced during respiration and taken up during photosynthesis

carbon fixation   the alternative name for the second stage of photosynthesis

carnivore   an animal that eats only other animals

cartilage   a tough material used to protect bones and strengthen the bronchi of the lungs

catalase   an enzyme which breaks down hydrogen peroxide to water and oxygen

catalyst   a substance that speeds up chemical reactions without changing itself

cell membrane a thin skin made of protein and fat which surrounds all cells

cell wall a tough outer coating of plant cells, made of fibres of cellulose

cellulose an insoluble carbohydrate composed of glucose molecules

cerebellum part of the brain that co-ordinates fine muscle control and balance

cerebrum part of the brain that controls conscious activities

chlorophyll a green pigment that traps sunlight for photosynthesis

chloroplast a tiny cell structure that contains chlorophyll

choice chamber a container in which animals are given a choice of environments

chromosome found in the nucleus of all cells, it is composed principally of DNA

cilia tiny cytoplasmic hairs which waft germs and dust from the lungs

circular muscle muscle surrounding a tube which, when it contracts, makes the bore of the tube smaller

CNS central nervous system composed of the brain and spinal cord

co-dominant alleles that have an equal effect on the phenotype of an organism

collecting duct the tube that collects urine from kidney nephrons, and where ADH acts

community all the organisms living in a habitat

competition the struggle between organisms for e.g. light, food, mates, territory

concentration gradient the gradient in concentration of molecules or atoms in a gas or liquid

consumer an organism that feeds upon other organisms; all animals are consumers

continuous variation variation in which there is a wide range of types

control an experimental set-up to demonstrate that a variable is having an effect

coronary artery the artery that supplies the heart muscle with blood from the aorta

cytoplasm the contents of a cell, excluding the nucleus

deamination the breakdown of excess amino acids with the production of urea

decomposer an organism that consumes dead organisms, causing decay

dehydration the removal of water from an organism or from food

denature the change in shape of an enzyme (protein) as a result of heat or change in pH

discontinuous variation variation in which there are only a few distinctly different types

diffusion the random spread of molecules or atoms down a concentration gradient

digestion the breakdown of complex insoluble molecules to soluble products

DNA deoxyribonucleic acid; the molecule that carries the genetic code

dominant an allele that always shows itself in the heterozygous genotype

ecosystem a habitat along with the community of animals and plants living in it

effector a muscle or gland that responds to nervous stimulation

environment the surroundings that influence an organism

enzyme a protein that acts as a catalyst in metabolic reactions

epithelium lining tissue in animals, e.g. cheek cells, ciliated epithelium of lungs

ethanol (alcohol) the end product of anaerobic respiration

excretion the removal of waste products from cells or the bodies of organisms

F_1 and F_2 the shorthand symbols for the first and second generations in genetics

fat (a lipid) a compound composed of fatty acids and glycerol and the elements C, H and O

fatty acid a sub-unit of a fat molecule; there are normally three in one fat molecule

fermentation anaerobic breakdown of sugars by e.g. yeasts to produce alcohol

fertilisation the process of fusion of the nuclei of male and female gametes (sex cells)

filtration (kidney) the first stage in the treatment of blood, taking place in the glomerulus

flaccid a description of plant cells that are short of water and soft and limp

food chain a sequence of organisms that feed on one another; producer to consumer

food web a network of interlinked food chains in an ecosystem

fungus a plant-like organism that feeds as a decomposer; it cannot photosynthesise

gall bladder an organ that stores bile

gamete a sex cell, e.g. sperm and eggs; they contain only one set of chromosomes.

gasohol ethanol (alcohol) mixed with petrol to form a 'green' fuel

gastric juice the liquid produced by the stomach walls, to aid digestion

gene a section of chromosome or DNA that codes for a single protein

genetic engineering the process of transferring genes from one organism to a different organism

genetics the study of inheritance

genotype the genes or alleles an organism possesses; symbolised by letters of alphabet

glomerulus (kidney) a bundle of capillaries surrounded by the Bowman's capsule of a nephron

glucose a simple 6-carbon sugar that is the principal respiratory substrate

glycerol a sub-unit of a fat molecule

glycolysis the first stage of respiration in which glucose is converted to pyruvic acid

growth hormone a pituitary hormone that stimulates growth of the body

habitat the place where an organism lives

haemoglobin the red protein of blood cells that carries oxygen

heart (mammalian) a pump with four chambers that pumps blood around the body

hepatic artery the artery that supplies the liver with oxygenated blood

hepatic portal vein the vein that carries digested food from the gut to the liver

hepatic vein the vein that carries blood from the liver to the vena cava

herbivore an animal that eats only plant material

heterozygous having different alleles of a gene, e.g. in pea plants Tt

homozygous having identical alleles of a gene, e.g. in pea plants TT or tt

horticulture the science or art of growing garden plants

humidity a measure of the amount of water vapour (moisture) in the air

hypertonic a solution that has more dissolved substances in it than another solution

hypotonic a solution that has less dissolved substances in it than another solution

hypothalamus an organ linked to the pituitary that monitors blood temperature and composition

hypothesis a prediction based on observations, which can be tested by experimentation

insulin a pancreatic hormone that stimulates the removal of glucose from the blood

iodine solution used to test for starch in food

isotonic a term used to describe two solutions of the same concentration

kidney tubules tiny tubules of the kidney, involved in the reabsorbtion of useful substances

lacteal the central tube of a villus, which absorbs digested fats. Part of lymph system

lactic acid an acid produced by muscles and by bacteria in anaerobic conditions

lactose milk sugar; some bacteria can turn it into lactic acid and cause souring in milk

large intestine (colon) the last part of the intestine where water is absorbed from digested food

limiting factor a factor that limits the rate of a process, e.g. light can limit photosynthesis

lipase an enzyme that digests fats (lipids) to fatty acids and glycerol

liver   an organ which produces bile and carries out many other important metabolic processes

longitudinal muscle   muscle which, when it contracts, makes the bore of a tube wider

lymphocyte   a white blood cell that produces antibodies to fight infection

macrophage   a white blood cell that engulfs foreign particles by phagocytosis

maltose   the sugar produced when starch is digested by amylase

mammal   a warm-blooded vertebrate (backboned animal) with hair or fur

marine bony fish   a fish that lives in the sea and has a bony skeleton

medulla (oblongata)   part of brain that controls basic bodily functions such as heart rate

meiosis   nuclear division that halves the chromosome number in the formation of gametes

mesenteric artery   the artery that carries blood from the aorta to the gut

mesophyll   plant cells between the epidermis cells of leaves, which contain chloroplasts

metabolic water   water produced as a result of aerobic respiration and other chemical reactions

metabolism   all the chemical reactions which take place in an organism

methane   the principle gas found in biogas

monohybrid cross   a genetics cross where only one characteristic is studied

motor neurone   a nerve cell that carries impulses from the brain to a muscle or gland

motor strip   the part of the cerebrum where motor neurones originate

mucus   a slime produced by various organs of the body for protection and lubrication

natural selection   the selection by nature of the best adapted organisms for reproduction

negative feedback   a process in which the increase or decrease in a factor triggers its control

nephron   a microscopic filtering unit of a kidney

neurone   a nerve cell, e.g. a sensory, relay or motor neurone

niche   the position an organism occupies in an ecosystem with respect to other organisms

nitrogen   a gas found in the atmosphere (79%) and a component of protein molecules

nitrogenous waste   waste containing the element nitrogen

nucleus   the control centre of a cell, which contains DNA in the form of chromosomes

oesophagus   the tube that carries food and drink from the mouth to the stomach

optimum   the best or ideal conditions

osmoreceptor   an organ that detects changes in the osmotic concentration of fluids

osmoregulation   the regulation of the concentration of fluids such as blood plasma

osmosis   the diffusion of water molecules through a selectively-permeable membrane

ovary   an organ of flowers and animals in which eggs (ova) are produced

oxygen   a gas found in the atmosphere (20%) which is produced during photosynthesis

oxygen debt   the oxygen required to remove lactic acid from the body after exercise

oxyhaemoglobin   haemoglobin, when it is carrying oxygen

pancreas   an organ that produces enzymes for digestion and some hormones

pepsin   an enzyme manufactured in the stomach, which digests protein to polypeptides

peristalsis   waves of muscular movements which, for example, push food along the gut

permeable   that which permits substances (often water) to pass through it

pH   a measure of the acidity or alkalinity of a fluid on a scale of 1 to 14

phagocytosis   the process of engulfing, carried out by macrophages

phenotype   the appearance brought about by the genes of an organism

phosphorylase   an enzyme that catalyses the conversion of glucose into starch

photolysis the splitting of water using light energy in the first stage of photosynthesis

photosynthesis the manufacture of sugars from water and CO_2 using light energy

pituitary a gland attached to the underside of the brain which produces hormones

plasma the fluid portion of the blood

plasmid a ring of DNA found in bacteria which is used in genetic engineering

plasmolysis the shrinkage of the cell contents of plant cells placed in hypertonic solutions

pollen (nucleus) the male gamete of flowering plants; produced in the anthers

polygenic inheritance the inheritance of characteristics controlled by many genes

population the number of individuals of one species of organism in an ecosystem

predator an animal that hunts and kills other animals called its prey

prey an animal that is hunted and killed by other animals called its predators

primary consumer an animal at the second stage of a food chain, which eats plants

producer an organism at the beginning of a food chain, which makes food

protease an enzyme that digests protein to peptides or amino acids

protein a compound composed of carbon, hydrogen, oxygen and nitrogen, required for growth and repair and many other functions

pulmonary artery the artery that carries deoxygenated blood from the heart to the lungs

pulmonary vein the vein that carries oxygenated blood from the lungs to the heart

pulse the rhythmic movement of the arteries due to the pumping action of the heart

pyramid of biomass a diagram to show the total mass of all organisms at each stage in a food chain

pyramid of energy as above, but showing the total energy stored at each stage

pyramid of numbers as above, but showing the total numbers of organism at each stage

pyruvic acid the 3-carbon compound that is the end-product of glycolysis

random assortment the random alignment of matching pairs of chromosomes during meiosis

reabsorbtion (kidney) the process of taking back useful substances from the filtrate

reagent a chemical substance used to detect another substance

receptor a nerve ending that receives information

recessive an allele that has no effect on phenotype unless with another of the same

rectum the last part of the large intestine, where faeces are stored

red blood cell a small cell that is packed with haemoglobin to transport oxygen

reflex action an automatic, quick response to a stimulus, involving a reflex arc

reflex arc a nerve pathway from sensory neurone to motor neurone via a relay neurone

relay neurone a nerve cell that links a sensory with a motor neurone, making a reflex arc

renal artery the artery that supplies the kidneys with blood from the aorta

renal vein the vein that takes blood from the kidneys to the vena cava

respiration the chemical process in cells, which releases energy (ATP) from food

saliva a digestive juice produced in the mouth, containing mucus and amylase

secondary consumer an animal at the third stage of a food chain, which eats other animals

selective breeding choosing animals and plants with useful characteristics for breeding purposes

selectively-permeable a description of membranes that allow some substances to pass through them

sensory neurone a neurone that receives information from the internal/external environment

sensory strip the part of the cerebrum where sensory neurones terminate

Exam skills and support

small intestine   the part of the gut where digestion is completed and foodstuffs are absorbed

species   a group of organisms capable of interbreeding to produce fertile offspring

specificity   a description of the precise nature of the action of enzymes and antibodies

spinal cord   the part of the CNS that links the brain to the rest of the body

starch   a storage carbohydrate in plants, composed of glucose molecules

stimulus   something that arouses a response from an animal or plant

stomach   an organ containing hydrochloric acid, and pepsin for the digestion of protein

substrate   (enzymes) the substance on which an enzyme acts

synthesis reaction   a reaction in which complex substances are made from simple substances

testes   male sex organs that manufacture sperm

tertiary consumer   the third consumer in a food chain

tissue fluid   fluid which surrounds the cells of the body

trachea   the main breathing tube from the mouth to the lungs

true-breeding   producing offspring that are genetically identical to the parent

turgid   a description of plant cells that are full of water and firm

urea   a nitrogenous waste product, produced by the liver and removed by the kidneys

ureter   the tube leading from a kidney to the bladder

urethra   the tube leading from the bladder to the outside

vacuole   a bag of fluid found in many plant cells, which gives the cell firmness

valve   found in veins and the heart to stop blood flowing backwards

variable   a factor in an experiment that may have to be controlled

vasoconstriction   the constriction of blood vessels in the skin to save heat

vasodilation   the expansion of blood vessels in the skin to increase heat loss

vein   a blood vessel that carries blood away from an organ, towards the heart

vena cava   the principal vein of the body, which takes blood to the heart

ventricle   a lower chamber of the heart, which pumps blood to the body via an artery

villus   a tiny finger-like projection in the gut which absorbs digested food

vitamin   an essential part of the diet; needed for the proper functioning of enzymes

X-chromosome   a sex chromosome, two of which determines the female sex (XX)

Y-chromosome   a sex chromosome, one of which determines the male sex (XY)

yeast   a single-celled fungus used in baking and brewing to produce CO_2 and ethanol

yoghurt   a milk-product resulting from the fermentation activity of bacteria

zygote   a fertilised egg

Top Tip Don't forget the flash cards!

Top Tip Best of luck in the exams!